T0300119

Interaction of Coronavirus Disease 2019 with other Infectious and Systemic Diseases

The book *Interaction of Coronavirus Disease 2019 with other Infectious and Systemic Diseases* elaborates on the pathogenesis, co-infection and association of COVID-19 with other infectious and systemic diseases. Moreover, this book explains pharmacological interventions and therapeutic targets in COVID-19. All chapters cover the different aspects of COVID-19. This book gives firsthand information on the co-infection, post-COVID effects, nanomedicine and pharmacogenomics in COVID-19.

The book offers comprehensive coverage of the most essential topics, including:

- Recent developments in COVID-19 pathogenesis and treatment
- Association of COVID-19 with infectious diseases and systemic diseases
- COVID-19 and co-infection
- Post-COVID and long-term COVID problems
- Nanomedicine and pharmacogenomics in COVID-19

While other books may touch on COVID-19 pathophysiology and treatment strategies, the unique aspect of this book is that it covers post-COVID effects and interaction of coronavirus disease 2019 with other infectious and systemic diseases.

M. Tabish Qidwai is M. Tech and PhD graduate of the Biotechnology Engineering at the Dr. A.P.J. Abdul Kalam Technical University Lucknow, India, and an associate professor in the SRM University, UP, India. He previously worked as Officiating Director of IBST, SRM University, Assistant Professor at BB Ambedkar University, Lucknow. He has been working in the area of infectious disease for more than one decade. He works as section editor of Current Indian Science (Bentham Science). He has been awarded several prestigious awards and appreciation letters.

Interaction of Coronavirus Disease 2019 with other Infectious and Systemic Diseases

Edited by
Dr. Tabish Qidwai
SRM University, India

CRC Press is an imprint of the
Taylor & Francis Group, an **informa** business

Designed cover image:

First edition published 2024
by CRC Press
6000 Broken Sound Parkway NW, Suite 300, Boca Raton, FL 33487–2742

and by CRC Press
4 Park Square, Milton Park, Abingdon, Oxon, OX14 4RN

CRC Press is an imprint of Taylor & Francis Group, LLC

Library of Congress Cataloging-in-Publication Data
Names: Qidwai, Tabish, editor.
Title: Interaction of Coronavirus disease 2019 with other infectious and systemic diseases / edited by Dr. M. Tabish Qidwai, SRM University, India.
Description: First edition. | Boca Raton : CRC Press, 2024. | Includes bibliographical references and index.
Identifiers: LCCN 2023014027 (print) | LCCN 2023014028 (ebook) | ISBN 9781032350165 (hardback) | ISBN 9781032350233 (paperback) | ISBN 9781003324911 (ebook)
Subjects: LCSH: COVID-19 (Disease) | Coronavirus infections. | COVID-19 (Disease)—Complications. | COVID-19 (Disease)—Pathogenesis.
Classification: LCC RA644.C67 I5786 2024 (print) | LCC RA644.C67 (ebook) | DDC 362.1962414—dc23/eng/20230527
LC record available at https://lccn.loc.gov/2023014027
LC ebook record available at https://lccn.loc.gov/2023014028

ISBN: 978-1-032-35016-5 (hbk)
ISBN: 978-1-032-35023-3 (pbk)
ISBN: 978-1-003-32491-1 (ebk)

DOI: 10.1201/9781003324911

Typeset in Times
by Apex CoVantage, LLC

Contents

PART II SARS CoV-2 Co-Infection with Other Pathogens

PART IV COVID-19 and Cancer

PART V Current Therapeutic Approaches

Contributors

Aashna Srivastava
School of Bio-Engineering and Food Technology, Shoolini University, Himachal Pradesh, India

Amreen Shamsad
Molecular & Human Genetics Laboratory, Department of Zoology, University of Lucknow, Lucknow, India

Anand Kumar Maurya
Department of Microbiology, All India Institute of Medical Sciences (AIIMS) Bhopal, Madhya Pradesh, India

Anand Prakash
Department of Biotechnology, Mahatma Gandhi Central University Bihar, Motihari, India

Ankit Madeshiya
Division of Toxicology and Experimental Medicine, CSIR-Central Drug Research Institute, Lucknow, UP, India

Atar Singh Kushwah
Molecular & Human Genetics Laboratory, Department of Zoology, University of Lucknow, Lucknow, India
and
Icahn School of Medicine at Mount Sinai, New York, NY, USA

Bechan Sharma
Department of Biochemistry, University of Allahabad, Allahabad, India

Bhartendu Nath Mishra
Department of Biotechnology, Institute of Engineering and Technology, Sitapur Road
Lucknow, India

Divya Chaudhary
Department of Pharmaceutical Sciences, Babasaheb Bhimrao Ambedkar University, Vidya Vihar, Raebareli Road, Lucknow, (U.P.), India

Hanan A. Balto
Department of Restorative Dental Sciences, College of Dentistry, King Saud University, Riyadh, Saudi Arabia

Hemant Kumar Bid
CP Labs, Cannabis Testing Laboratory, Columbus, Ohio, USA

Krishan Kumar
Department of Pharmaceutical Engineering & Technology, Indian Institute of Technology (Banaras Hindu University), Varanasi, (U.P.), India

Tabish Qidwai
Faculty of Biotechnology, Shri Ramswaroop Memorial University, Lucknow, Uttar Pradesh, India

Malti Dadheech
Department of Microbiology, All India Institute of Medical Sciences (AIIMS) Bhopal, Madhya Pradesh, India

Mariam Imam
The Norwegian College of Fishery Science, Faculty of Biosciences, Fisheries & Economics, University of Tromsø, The Arctic University of Norway, Tromsø, Norway

Mohini Mishra
Department of Pharmaceutical Engineering & Technology, Indian Institute of Technology (Banaras Hindu University), Varanasi, (U.P.), India

Monisha Banerjee
Molecular & Human Genetics Laboratory, Department of Zoology, University of Lucknow, Lucknow, India

Neha Kapoor
Department of Chemistry, Hindu College, University of Delhi, Delhi, India

Nikhat J. Siddiqi
Biochemistry Department, College of Science, King Saud University, Riyadh

Osaid Masood
Molecular & Human Genetics Laboratory, Department of Zoology, University of Lucknow, Lucknow, India

Reem Hamoud Alrashoudi
Department of Clinical Laboratory Sciences, College of Applied Medical Sciences, King Saud University, Riyadh, Saudi Arabia

Rishabh Chaudhary
Department of Pharmaceutical Sciences, Babasaheb Bhimrao Ambedkar University, Vidya Vihar, Raebareli Road, Lucknow, (U.P.), India

Ruchi Chawla
Department of Pharmaceutical Engineering & Technology, Indian Institute of Technology (Banaras Hindu University), Varanasi, (U.P.), India

Sabiha Fatima
Department of Clinical Laboratory Sciences, College of Applied Medical Sciences, King Saud University, Riyadh, Saudi Arabia

Samina Wasi
Department of Biochemistry, College of medicine, Imam Abdulrahman Bin Faisal University, Alkhobar, Saudi Arabia

Sanjay Saini
Department of Physiology and Cellular Biophysics, Columbia University Irving Medical Center, New York, NY, USA

Sanjay Singh
Department of Biochemistry, Microbiology, and Bioinformatics, Faculty of Science and Engineering, Université Laval Québec, QC, Canada G1V 4G2

Sarika Singh
Division of Toxicology and Experimental Medicine,

CSIR-Central Drug Research Institute, Lucknow, UP, India
and
Academy of Scientific & Innovative Research (AcSIR), Ghaziabad, India

Saurabh Kumar
Molecular & Human Genetics Laboratory, Department of Zoology, University of Lucknow, Lucknow, India

Shama Parveen
Molecular & Human Genetics Laboratory, Department of Zoology, University of Lucknow, Lucknow, India

Shilpa Kumari
Division of Toxicology and Experimental Medicine, CSIR-Central Drug Research Institute, Lucknow, UP, India

Shireen Masood
Molecular & Human Genetics Laboratory, Department of Zoology, University of Lucknow, Lucknow, India

Sooad K. AlDaihan
Biochemistry Department, College of Science, King Saud University, Riyadh

Sukanya Tripathy
Molecular & Human Genetics Laboratory, Department of Zoology, University of Lucknow, Lucknow, India
and
Department of Biotechnology, Babasaheb Bhimrao Ambedkar University, Lucknow, India

Tejal Shreeya
Institute of Biophysics, Biological Research Centre, Szeged, Hungary, Europe
and
Doctoral School of Theoretical Medicine, University of Szeged, Hungary, Europe

Varsha Rani
Department of Pharmaceutical Engineering & Technology, Indian Institute of Technology (Banaras Hindu University), Varanasi, (U.P.), India

Vikas Mishra
Department of Pharmaceutical Sciences, Babasaheb Bhimrao Ambedkar University, Vidya Vihar, Raebareli Road, Lucknow, (U.P.), India

Part I

Current Trends in Pathophysiology and Post-COVID Effects

1 Pathophysiology of SARS-CoV-2

Tejal Shreeya, Tabish Qidwai and Bhartendu Nath Mishra

1.1 INTRODUCTION

The coronavirus disease 2019 (COVID-19) is caused by the virus SARS-CoV-2. The SARS-CoV-2 (Severe Acute Respiratory syndrome coronavirus-2) virus originated in Wuhan, Hubei Province, China in late 2019 and caused a severe disease COVID-19. [1] The outbreak of this virus led to the emergency situation in the whole world, thereby being declared a pandemic. It has been spreading globally ever since, and it still poses constant threat to public health. [2]

Patients with comorbidities such as diabetes and hypertension are at a higher risk of COVID-19 infection.

Though it's a respiratory disease, it causes extra respiratory symptoms as well. Anosmia (loss of smell) and ageusia (loss of taste) are the characteristic symptoms of the disease. The other major symptoms of the disease are headache, high fever, difficulty in breathing, body ache, dry throat and cough. In severe cases it leads to death. [3]

The reproductive ratio (R0) is the rate of transmission of virus. For COVID-19, R0 is 2–3. This means a single COVID-19 positive patient can spread the disease to three other persons, and those infected individuals can spread it to another three people.

1.2 STRUCTURE OF THE VIRUS

The SARS-CoV-2 is a linear, single-stranded, positive-sense RNA coronavirus with approximately 30kb genome and belongs to the β-coronaviridae family. [4] It has 14 ORFs (Open Reading frames). The 5' end of the virus encodes for polyprotein and non-structural proteins while the 3' end is responsible for encoding the four structural proteins namely S (Spike protein), M (Membrane), E (Envelope) and N (Nucleocapsid). [5] The Spike protein is composed of two subunits S1 and S2. The S1 subunit is responsible for binding with the host cells receptors while the S2 subunit is responsible for the fusion of the virus and the host cell. The S1 subunit of the S protein recognises the receptor on the host cell and facilitates the attachment of the virus with the host cell. Apart from recognition and viral

DOI: 10.1201/9781003324911-2

attachment, priming of S protein is required for the viral entry into the host cell which is done by the serine protease TMPRSS2 and/or cathepsin B/L. [6]

The S1 subunit has the RBD which binds to the host cell receptor. This RBD has two different conformations. It is either present in the inactive state or down conformation, which does not allow binding to the receptors by creating a steric hindrance, or it is present in an active state or up conformation, where the RBD is open to bind with the receptor.

1.3 ENTRY OF SARS-CoV-2

The S1 subunit recognises the ACE2 receptor present on the host cells. The S1 and S2 subunits are separated by furin cleavage site, which is cleaved by protease present in the host cell. Meanwhile the S1 subunit recognises the ACE2, the transmembrane protease TMPRSS2 cleaves the S2 site which is another cleavage site present in the S2 subunit. The cleavage activates the S2 subunit to fuse viral membrane to the host cell membrane. [7–9]

The other route of entry is the endosome, in which the cathepsin cleaves the spike protein. [10, 11] The protease is highly present in the epithelial cells of respiratory, gastrointestinal, urogenital tract and brain cell lines. [12]

1.4 ENTRY FACTORS OF SARS-CoV2

ACE2 is the key receptor for the entry of SARS-CoV-2 in the cells. Apart from ACE2, NRP-1, AXL and Cd147 might be a putative co receptor and aid in the viral entry. TMPRSS2 and Cathepsin B/L are the protease which help in the entry of virus in the host cell by priming the Spike protein.

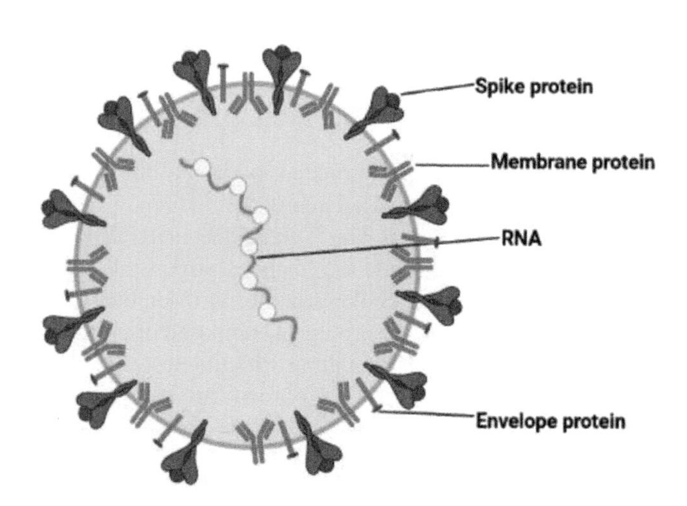

FIGURE 1.1 Structure of SARS-CoV-2 showing all the structural proteins.

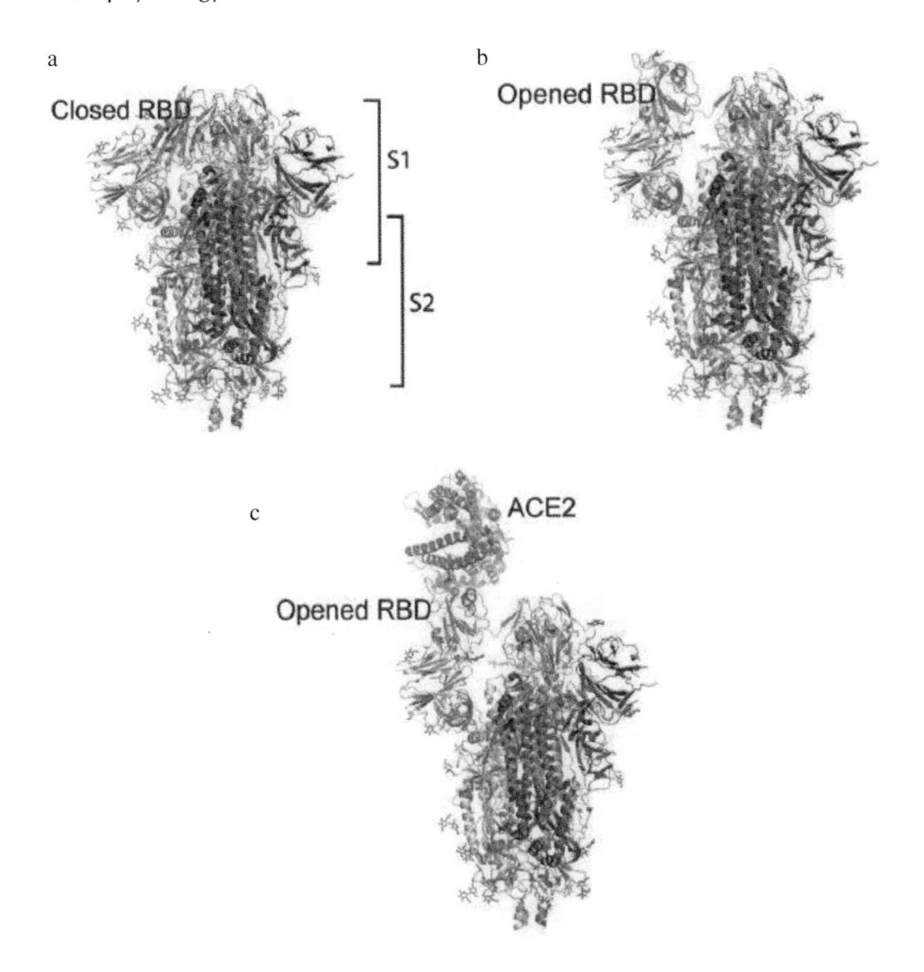

FIGURE 1.2 a) Structure of RBD in a closed conformation b) Structure of RBD in an open conformation and c) Open conformation of RBD showing binding with ACE2 receptor. [13]

1.5 HALLMARK OF COVID-19

The key receptor for host cell entry is ACE2, which is highly expressed in the respiratory tract. The olfactory mucosa that is present in the upper respiratory tract is in direct contact with the CNS via the olfactory neurons. The loss of smell which is the hallmark of the COVID-19 might invade CNS through olfactory mucosa via retrograde route. [14] The exact mechanism of entry of SARS-CoV-2 virus is the CNS is still elusive. The long-lasting COVID-19 symptoms that is the persistent loss of olfactory function in humans with SARS-CoV-2 is the result of direct damage to the olfactory sensory neurons that are responsible for detecting odour in the olfactory epithelium. [15]

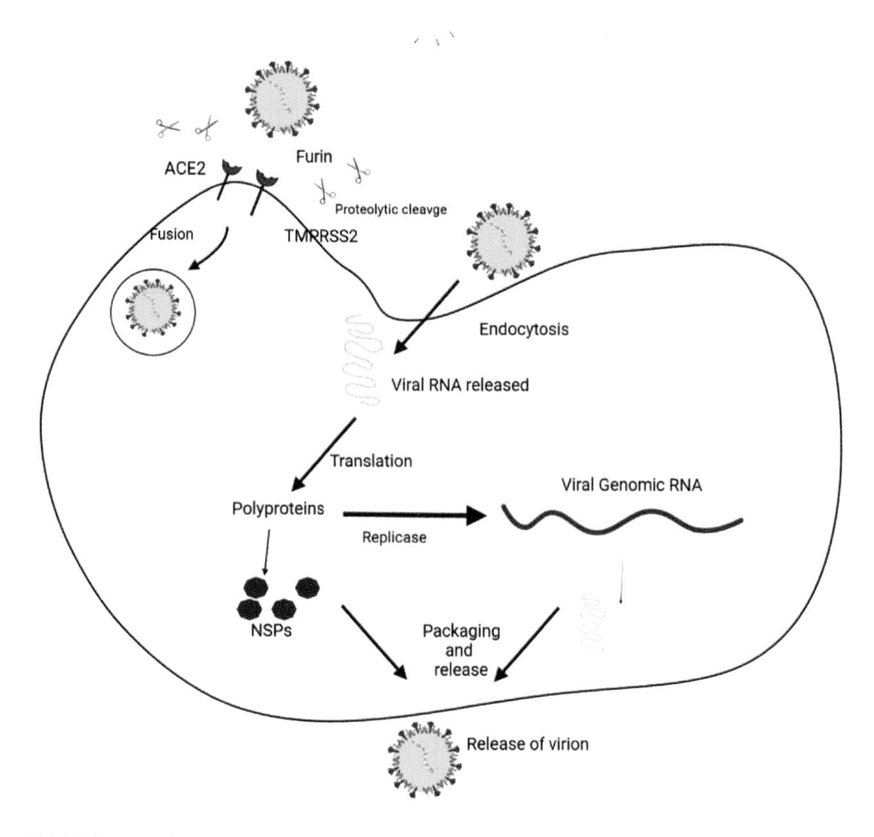

FIGURE 1.3 Showing the recognition of SARS-CoV-2 spike protein by ACE2 and its priming by protease enzymes.

1.5.1 Pathophysiology of SARS-CoV-2

The SARS-CoV-2 uses nasal pathway for its transmission (sneezing and coughing). The virus enters the lungs through the respiratory tract. Once inside the lungs, the spike protein of the virus recognises the ACE2 receptor present on the epithelial cells of lungs. Upon recognition, binding occurs followed by priming (by TMPRSS2) and entry into the host cell. Once inside the cell, the single-stranded RNA (ssRNA) is released by the virus in the cytoplasm. This ssRNA undergoes transcription and translation using host cells resources. ssRNA uses host cells ribosome to produces polyproteins. It also uses RNA-dependent RNA polymerase to duplicate its RNA. The packaging structures are also synthesised by the host cell. [7, 16, 17]

1.6 AFFECTING RESPIRATORY TRACT

The nasopharyngeal cavity and the respiratory tract have really high tropism for SARS-CoV-2 virus, due to high expression of ACE2 receptor. [18] The virus infects the respiratory tract in different stages. The nasopharyngeal cavity is the first to be affected, followed by bronchi and bronchioles, further followed by alveoli (pneumocytes). [19] Infection of pneumocytes, increases the expression of interferons and interleukins. This activates the immune cells of the alveoli such as macrophages and in turn recruits neutrophils. [20] The infected cells release inflammatory mediators such as interleukins which trigger a cascade of events such as increased pneumocyte damage, infiltration of plasma proteins. The alveoli further gets filled with dead cells and viral particles hence increasing the volume of interstitium between capillary and alveoli chamber. This results in a compromised gas exchange. This phenomenon gives the disease its name severe acute respiratory syndrome (SARS). [20, 21]

1.7 SARS-CoV-2 INFECTION AFFECTS RAAS

The coronavirus is known to be mediated by TMPRSS2 and ACE2. [7] The endothelial cells present in the lungs and respiratory tract highly express ACE2, thereby making a direct target of infection. The normal physiological process of ACE2 is involved in regulation and metabolism of renin-angiotensin-aldosterone system (RAAS). ACE2 converts Angiotensin II to angiotensin 1–7, hence regulating vasodilation. The spike protein of SARS-CoV-2 triggers the downregulation of ACE2 in the lung tissue, by directly binding to the ACE2 receptor. Since SARS-CoV-2 targets ACE2 receptor is not readily available, so higher binding of Ang II to AT1 receptor and ultimately causing vasoconstriction, which leads to enhanced inflammation and even thrombosis. [22]

1.8 ENDOTHELIAL DYSFUNCTION AND PLATELET ACTIVATION

The endothelial cells maintain the hemostasis. Endothelial cells when encounter PAMP's, proinflammatory cytokines and DAMP's provokes them to become activated. Endothelial activation contribute to platelet activation, increased vascular permeability. Activated endothelial cells leads to amplification of enzymatic activity of coagulation cascade protein, which triggers thrombin generation and clot formation.

The platelet activation plays an important role in the pathophysiology of COVID-19 disease. Either due to cytokine storm or endothelial activation, the platelet activation is intensified. Due to impaired platelet production and immune depletion, thrombocytopenia occurs which is another important hallmark of COVID-19 disease. [23]

1.9 ENDOCRINE DYSFUNCTION

Diabetes has been associated with increased SARS-CoV-2 disease associated with lethality and death. [24–26] It has been found that patients with Type II Diabetes have elevated ACE2 expression in bronchi and alveoli, which makes them more prone to the COVID-19 disease. [27] SARS-CoV-2 infection has been observed to cause ketoacidosis and increased blood glucose level due to immune dysregulation.

1.9.1 Neuropathology

Though COVID-19 is a respiratory disease, it is not just confined to the respiratory tract. Forty-five percent of the COVID-19 infected people had complaints with neurological complications. The neurotropic nature of this virus has led to neurological symptoms as well. [15]

It's presence has been found in CNS (viral RNA in CSF) as well. [28] Patients infected with COVID-19 disease were diagnosed with various neurological complications. It is still ambiguous whether it is a direct effect of SARS-CoV-2 or an indirect effect as a result of systemic inflammation. The acute and chronic neurological symptoms include headache, nausea, seizures, delirium, confusion, etc. It also has long-term complications such as persistent fatigue, meningitis and encephalitis.

Though the CNS is protected by BBB and B-CSFB, yet the virus breached these barriers to enter CNS. The possible routes of entry might be as follows: the SARS-CoV-2 might enter the brain via olfactory nerve. [29] The presence of viral RNA has been detected in CSF of many infected patients, which clearly indicates viral invasion in CNS via crossing BBB. [30] The other indirect route may be the shedding of SARS-CoV-2 protein and cytokines, which may in turn cross BBB and affect CNS function. [31, 32] SARS-CoV-2 might gain access into the neurons and astrocytes via NRP-1, as ACE2 is not expressed by these cells and affects neurons and epithelial cells of choroid plexus. [33, 34]

COVID-19 is characterised by systemic inflammation which might increase the permeability of the BBB, thereby allowing infected immune cells and cytokines to breach BBB and enter CNS. [35] Other than systemic inflammation, COVID-19 leads to cytokine storm, lymphopenia, thrombosis, etc. It also leads to various cerebrovascular diseases and strokes and also increases the risk of neurodegenerative disease such as AD (Alzheimer's disease).

1.10 LIST OF ABBREVIATIONS

SARS-CoV-2	Severe acute respiratory syndrome-2
ACE2	Angiotensin-converting enzyme
TMPRSS2	Transmembrane protease serine 2
RAAS	Renin-angiotensin-aldosterone System
B-CSFB	Blood cerebrospinal fluid barrier
ssRNA	Single-stranded ribonucleic acid
RBD	Receptor-binding domain
Ro	Reproductive ratio
NRP-1	Neuropilin-1

CNS	Central nervous system
CSF	Cerebrospinal fluid
BBB	Blood-brain barrier
RNA	Ribonucleic acid
PAMPs	Pathogen associate molecular patterns
DAMPs	Damage associated molecular pattern
Ang II	Angiotensin II

ACKNOWLEDGEMENT

We would like to thank our institutions Biological Research Centre, University of Szeged, Szeged, Hungary and SRM University for providing a supportive environment to carry out this work.

REFERENCES

1. S. Picot et al., "Coalition: Advocacy for prospective clinical trials to test the post-exposure potential of hydroxychloroquine against COVID-19," *One Heal.*, vol. 9, January 2020, doi: 10.1016/j.onehlt.2020.100131.
2. P. Zhou et al., "A pneumonia outbreak associated with a new coronavirus of probable bat origin," *Nature.*, vol. 579, no. 7798, pp. 270–273, 2020, doi: 10.1038/s41586-020-2012-7.
3. Y. Ding et al., "Organ distribution of severe acute respiratory syndrome (SARS) associated coronavirus (SARS-CoV) in SARS patients : Implications for pathogenesis and virus transmission pathways," pp. 622–630, 2004, doi: 10.1002/path.1560.
4. C. Li, Y. Yang, and L. Ren, "Genetic evolution analysis of 2019 novel coronavirus and coronavirus from other species," *Infect. Genet. Evol.*, vol. 82, March 2020, doi: 10.1016/j.meegid.2020.104285.
5. L. Chen, and L. Zhong, "Genomics functional analysis and drug screening of SARS-CoV-2," *Genes Dis.*, vol. 7, no. 4, pp. 542–550, 2020, doi: 10.1016/j.gendis.2020.04.002.
6. G. Simmons, D. N. Gosalia, A. J. Rennekamp, J. D. Reeves, S. L. Diamond, and P. Bates, "Inhibitors of cathepsin L prevent severe acute respiratory syndrome coronavirus entry," *Proc. Natl. Acad. Sci. U. S. A.*, vol. 102, no. 33, pp. 11876–11881, 2005, doi: 10.1073/pnas.0505577102.
7. "SARS-CoV-2 cell entry depends on ACE2 and TMPRSS2 and is blocked by a clinically proven protease inhibitor," *Cell*, vol. 181, no. 2, 2020, doi: 10.1016/j.cell.2020.02.052.
8. J. Beumer et al., "A CRISPR/Cas9 genetically engineered organoid biobank reveals essential host factors for coronaviruses," *Nat. Commun.*, vol. 12, no. 1, 2021, doi: 10.1038/s41467-021-25729-7.
9. H. J. Solinski, T. Gudermann, and A. Breit, "Pharmacology and signaling of MAS-related G protein-coupled receptors," *Pharmacol. Rev.*, vol. 66, no. 3, pp. 570–597, 2014, doi: 10.1124/pr.113.008425.

10. M. M. Lamers et al., "Human airway cells prevent sars-cov-2 multibasic cleavage site cell culture adaptation," *Elife.*, vol. 10, pp. 1–22, 2021, doi: 10.7554/ELIFE.66815.
11. A. Z. Mykytyn et al., "Sars-cov-2 entry into human airway organoids is serine protease-mediated and facilitated by the multibasic cleavage site," *Elife.*, vol. 10, pp. 1–23, 2021, doi: 10.7554/ELIFE.64508.
12. J. Qiao et al., "The expression of SARS-CoV-2 receptor ACE2 and CD147, and protease TMPRSS2 in human and mouse brain cells and mouse brain tissues," *Biochem. Biophys. Res. Commun.*, vol. 533, no. 4, pp. 867–871, 2020, doi: 10.1016/j.bbrc.2020.09.042.
13. Y. Huang, C. Yang, X. feng Xu, W. Xu, and S. wen Liu, "Structural and functional properties of SARS-CoV-2 spike protein: Potential antivirus drug development for COVID-19," *Acta Pharmacol. Sin.*, vol. 41, no. 9, pp. 1141–1149, 2020, doi: 10.1038/s41401-020-0485-4.
14. M. S. Lener, "乳鼠心肌提取 HHS public access," *Physiol. Behav.*, vol. 176, no. 1, pp. 139–148, 2016, doi: 10.1021/acschemneuro.6b00043.The.
15. G. D. de Melo et al., "COVID-19-related anosmia is associated with viral persistence and inflammation in human olfactory epithelium and brain infection in hamsters," *Sci. Transl. Med.*, vol. 13, no. 596, pp. 1–25, 2021, doi: 10.1126/scitranslmed.abf8396.
16. C. J. A. Sigrist, A. Bridge, and P. Le Mercier, "A potential role for integrins in host cell entry by SARS-CoV-2," *Antiviral Res.*, vol. 177, p. 104759, March 2020, doi: 10.1016/j.antiviral.2020.104759.
17. J. Vallamkondu et al., "SARS-CoV-2 pathophysiology and assessment of coronaviruses in CNS diseases with a focus on therapeutic targets," *Biochim. Biophys. Acta—Mol. Basis Dis.*, vol. 1866, no. 10, p. 165889, 2020, doi: 10.1016/j.bbadis.2020.165889.
18. M. Li, L. Li, Y. Zhang, and X. Wang, "An investigation of the expression of 2019 novel coronavirus cell receptor gene ACE2 in a wide variety of human tissues," *Infect. Dis. Poverty.*, vol. 9(1), no. 45, pp. 1–7, 2020 [Online]. https://doi.org/.
19. R. J. Mason, "Thoughts on the alveolar phase of covid-19," *Am. J. Physiol.—Lung Cell. Mol. Physiol.*, vol. 319, no. 1, pp. L115–L120, 2020, doi: 10.1152/ajplung.00126.2020.
20. R. Alon et al., "Leukocyte trafficking to the lungs and beyond: Lessons from influenza for COVID-19," *Nat. Rev. Immunol.*, vol. 21, no. 1, pp. 49–64, 2021, doi: 10.1038/s41577-020-00470-2.
21. J. G. Bartlett, "The severe acute respiratory syndrome," *Infect. Dis. Clin. Pract.*, vol. 12, no. 3, pp. 218–219, 2004, doi: 10.1097/01.idc.0000129853.80250.2c.
22. P. Verdecchia, C. Cavallini, A. Spanevello, and F. Angeli, "The pivotal link between ACE2 deficiency and SARS-CoV-2 infection," *Eur. J. Intern. Med.*, vol. 76, April, pp. 14–20, 2020, doi: 10.1016/j.ejim.2020.04.037.
23. S. Q. Jiang, Q. F. Huang, W. M. Xie, C. Lv, and X. Q. Quan, "The association between severe COVID-19 and low platelet count: Evidence from 31 observational studies involving 7613 participants," *Br. J. Haematol.*, vol. 190, no. 1, pp. e29–e33, 2020, doi: 10.1111/bjh.16817.
24. N. Stefan, A. L. Birkenfeld, and M. B. Schulze, "Global pandemics interconnected—obesity, impaired metabolic health and COVID-19," *Nat. Rev. Endocrinol.*, vol. 17, no. 3, pp. 135–149, 2021, doi: 10.1038/s41574-020-00462-1.
25. S. A. Bae, S. R. Kim, M. N. Kim, W. J. Shim, and S. M. Park, "Impact of cardiovascular disease and risk factors on fatal outcomes in patients with COVID-19 according to age: A systematic review and meta-analysis," *Heart.*, vol. 107, no. 5, pp. 373–380, 2021, doi: 10.1136/heartjnl-2020-317901.

26. J. Wang, and W. Meng, "COVID-19 and diabetes: The contributions of hyperglycemia," *J. Mol. Cell Biol.*, vol. 12, no. 12, pp. 958–962, 2020, doi: 10.1093/jmcb/mjaa054.
27. S. R. A. Wijnant et al., "Expression of ACE2, the SARS-CoV-2 receptor, in lung tissue of patients with type 2 diabetes," *Diabetes.*, vol. 69, no. 12, pp. 2691–2699, 2020, doi: 10.2337/db20-0669.
28. L. Pellegrini et al., "Short article SARS-CoV-2 infects the brain choroid plexus and disrupts the blood-CSF barrier in human brain organoids," *Stem Cell*, vol. 27, no. 6, pp. 951–961, 2020, doi: 10.1016/j.stem.2020.10.001.
29. J. Meinhardt et al., "Olfactory transmucosal SARS-CoV-2 invasion as a port of central nervous system entry in individuals with COVID-19," *Nat. Neurosci.*, vol. 24, no. 2, pp. 168–175, 2021, doi: 10.1038/s41593-020-00758-5.
30. T. Moriguchi et al., "A first case of meningitis/encephalitis associated with SARS-Coronavirus-2," *Int. J. Infect. Dis.*, vol. 94, pp. 55–58, 2020, doi: 10.1016/j.ijid.2020.03.062.
31. E. M. Rhea et al., "The S1 protein of SARS-CoV-2 crosses the blood–brain barrier in mice," *Nat. Neurosci.*, vol. 24, no. 3, pp. 368–378, 2021, doi: 10.1038/s41593-020-00771-8.
32. Y. Atmaram, S. L. Preston, R. P. Jeyadhas, A. E. Lang, R. Hammamieh, and A. H. Clayton, "Blood–brain barrier: COVID-19, pandemics, and cytokine norms," *Innov. Clin. Neurosci.*, vol. 18, no. 1, pp. 321–323, 2021.
33. F. Jacob et al., "Human pluripotent stem cell-derived neural cells and brain organoids reveal SARS-CoV-2 neurotropism predominates in choroid plexus epithelium," *Cell Stem Cell.*, vol. 27, no. 6, pp. 937–950, 2020, doi: 10.1016/j.stem.2020.09.016.
34. N. Chakravarty et al., "Neurological pathophysiology of SARS-CoV-2 and pandemic potential RNA viruses: A comparative analysis," *FEBS Lett.*, vol. 595, no. 23, pp. 2854–2871, 2021, doi: 10.1002/1873-3468.14227.
35. R. Sankowski, S. Mader, and S. I. Valdés-Ferrer, "Systemic inflammation and the brain: Novel roles of genetic, molecular, and environmental cues as drivers of neurodegeneration," *Front. Cell. Neurosci.*, vol. 9, pp. 1–20, 2015, doi: 10.3389/fncel.2015.00028.

2 Investigation of Post-COVID/Long COVID Complications in Humans

Sukanya Tripathy, Sanjay Singh, Monisha Banerjee and Anand Prakash

2.1 INTRODUCTION

The coronavirus responsible for severe acute respiratory syndrome (SARS-CoV-2) and maybe COVID-19 has spread over the globe. Despite vaccination efforts, SARS-CoV-2 has already infected over 1.1 million people in the United States alone, and novel variations like the Delta strain are making the disease even more contagious than before. [1–3] Although the elderly population had previously been where mortality rates were highest, the spread of the disease shifted to a younger, unvaccinated demographic as the susceptible population became more widely immunised. [4] Clinical introduction of COVID-19 has been widely variable, with respiratory complications frequently playing a significant role. The prevalence of SARS-CoV-2 is at an all-time high because of the severe long-term consequences the virus has caused for some infected individuals. Respiratory complications are a common feature of the clinical presentation of COVID-19, but this is not always the case.

Several people infected with SARS-CoV-2 went on to develop severe, life-threatening complications from the disease (Figure 2.1). Long-term COVID-19 has come to speak to wide complications and sequelae of indications that will emerge months after initial contamination, beyond initial reports of patients feeling exhausted for months. [5] Lung fibrosis, venous thromboembolism (VTE), blood vessel thromboses, cardiac thrombosis and aggravation, stroke, brain fog, dermatological complications and general temperament dysfunctions have all been considered as possible late complications of COVID-19 disease based on previous research. [6] Although the range of these complications is broad, certain patient characteristics have been shown to be predictive of which symptoms they will develop and for how long. [7] Here, we evaluate the long-term effects of COVID-19 and its pathophysiology. We also discuss some of the most important complications that can arise, including cardiovascular, neurological and mental, haematological, aspiratory, dermatological and other injuries.

DOI: 10.1201/9781003324911-3

2.2 PATHOPHYSIOLOGY

Several pathophysiological pathways of the virus may account for COVID-19's longer-term problems and sequelae, but the precise mechanisms responsible for these consequences are yet unclear. Direct viral tissue damage is a potential pathophysiological mechanism; the SARS-CoV-2 entry receptor, angiotensin-converting enzyme 2 (ACE2), is expressed in many different tissues and organs, allowing the virus to infect target cells by activating its spike protein with transmembrane serine protease. [8, 9] Since these receptors are found in epithelial cells, nasal goblet cells, gastrointestinal epithelial cells, pancreatic cells and renal podocytes, direct tissue destruction may be a fundamental mechanism of the presentation of SARS-CoV-2 infection, which may also contribute to its longer-term consequences. [10–12] Endothelial cells were shown to have strong ACE2 expression in early pandemic studies, and COVID-19 infection was found to significantly compromise the integrity of the vascular barrier and increase the procoagulative state. [13] Follow-up investigations of COVID-19 survivors have revealed the long-term effects of these alterations, with 71% of patients developing lung radiological abnormalities and 25% developing functional abnormalities after infection. [14]

There are other potential pathophysiological processes that may explain how COVID-19 causes a multiorgan systemic disease, not only direct cellular infection. Long-term COVID-19 infection problems have been linked to mechanisms, including endothelial damage, immune system dysregulation and hypercoagulability, which can result in thrombosis. Autoreactive T cells have been found in autopsies of deceased COVID-19-infected individuals, suggesting immune system dysregulation similar to that seen in autoimmune disease. [15]

Long-term complications from SARS-CoV-1, the virus that emerged before SARS-CoV-2 in 2003, have been studied extensively and found to be very similar to those from SARS-CoV-2. [16] The fact that they both use the ACE2 receptor to infect cells suggests that they use similar mechanisms of cell entry; however, SARS-CoV-2 has a higher affinity for the receptor and an extra cleavage site, which may allow for more efficient infection and possibly more severe longer-term complications. [17]

Patients with long-lasting symptoms of COVID-19 have either no detectable SARS-CoV-2 infection at all (PCR negative) or chronic low-level detection of SARS-CoV-2 (PCR positive) for traces of the virus. A Danish investigation of patients with post-symptomatic COVID-19 indicated that in cases of persistent PCR-positive infection, a SARS-CoV-2-specific CD8 T cell response of higher breadth and size may explain longer-lasting symptoms.

2.3 RESPIRATORY COMPLICATIONS

Several pathophysiological pathways of the virus may account for COVID-19 longer-term problems and sequelae, but the precise mechanisms responsible for these consequences are yet unclear. Direct viral tissue damage is a

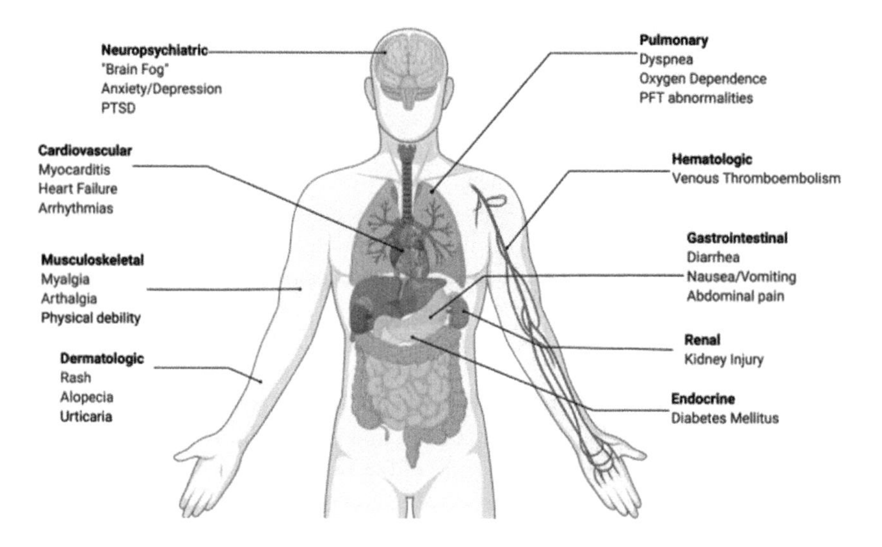

FIGURE 2.1 Parts of body affected by long-term COVID-complications.

potential pathophysiological mechanism; the SARS-CoV-2 entry receptor, angiotensin-converting enzyme 2 (ACE2), is expressed in many different tissues and organs, allowing the virus to infect target cells by activating its spike protein with transmembrane serine protease 2. Direct tissue damage may be a primary mechanism of the presentation of SARS-CoV-2 infection, which may also contribute to its longer-term complications, as these receptors are expressed in epithelial cells, nasal goblet cells, gastrointestinal epithelial cells, pancreatic cells and renal podocytes. Endothelial cells were shown to have strong ACE2 expression in early pandemic studies, and COVID-19 infection was found to significantly compromise the integrity of the vascular barrier and increase the pro-coagulative state. Follow-up investigations of COVID-19 survivors have revealed the long-term effects of these alterations, with 71% of patients developing lung radiological abnormalities and 25% developing functional abnormalities after infection.

There are other potential pathophysiological processes that may explain how COVID-19 causes a multiorgan systemic disease, not only direct cellular infection. Endothelial injury, immune system dysregulation and hypercoagulability, often leading to thrombosis, are also proposed pathways leading to long-term COVID-19 infection complications. Autoreactive T cells have been found in autopsies of deceased COVID-19-infected individuals, suggesting immune system dysregulation similar to that seen in autoimmune disease.

Long-term complications from SARS-CoV-1, the virus that emerged before SARS-CoV-2 in 2003, have been studied extensively and found to be very similar to those from SARS-CoV-2. The fact that they both use the ACE2 receptor to infect cells suggests that they use similar mechanisms of cell entry; however, SARS-CoV-2 has a higher affinity for the receptor and an extra cleavage site, which may allow for more efficient infection and possibly more severe longer-term complications.

2.4 CARDIOVASCULAR COMPLEXITIES

After being released from the hospital following COVID-19, cardiac problems are a commonly reported complication. About 21% of patients experienced chest discomfort 60 days after being released from the hospital, as reported by Carfi et al. [18] At 60-day follow-up, palpitations have been reported in as many as 9% of patients. Long-term effects of SARS-CoV-2 infection have been recorded, including subjective cardiac symptoms, but there have also been various quantitative results. Postural tachycardia syndrome (POTS) has been linked to SARS-CoV-2 infection, and its prevalence is now being studied. [19] Acute infection can lead to some of COVID-19 long-term cardiovascular consequences. Sub-clinical myocarditis was observed by cardiac magnetic resonance imaging in as many as 2.3% of participants in a large study of young, healthy, competitive collegiate athletes. Furthermore, routine cardiac magnetic resonance imaging found persistent inflammation in 60% of patients and persistent elevation in high-sensitivity troponin T in as much as 71% of patients after discharge from hospital with COVID-19. [20] Due to the lack of comparable research in other respiratory disorders, the clinical significance of this discovery is unclear; nonetheless, it does appear to show that SARS-CoV-2 infection can have long-lasting consequences on the heart. High frequencies of aberrant findings on echocardiography have also been linked to the acute phase of COVID-19. [21]

Several potential pathways have been postulated to explain how SARS-CoV-2 causes cardiac damage. Histological investigation of the heart has revealed a number of abnormalities in early descriptive autopsy studies, including interstitial inflammatory infiltration, myocardial enlargement and necrosis and RT-PCR positive for SARS-CoV-2. [22] Direct viral entry occurs because SARS-CoV-2 has affinity for the ACE2 receptor on the surface of cells, as has been thoroughly characterised. Researchers used single-cell RNA sequencing to determine that 7.5% of myocytes express ACE2, making the muscle a vulnerable organ to direct viral damage. [23] SARS-CoV-2 viral RNA was detected in the myocardium of 61.5% of patients in autopsy examinations, supporting this theory. Increased cytokine expression was also seen in patients with the greatest viral load (>1,000 copies). Myocarditis is caused, in part, by cell-mediated toxicity by CD8+ T lymphocytes in response to viral antigen presented following virus entrance into the cell and subsequent cardiac inflammation. [24] These pathways contribute to the presentation of COVID-19's cardiovascular problems and consequences in both the long- and short-term.

The use of renin-angiotensin-aldosterone system (RAAS) inhibition due to the virus' affinity for ACE2 receptors have been allayed by randomised data showing no difference in outcomes between continuing and stopping RAAS inhibition in hospitalised patients with COVID-19. Therefore, current society guidelines advise that patients with COVID-19 should continue RAAS-related therapy as needed. [25] With their antithrombotic and anti-inflammatory qualities as a working hypothesis, statins have recently been studied for their potential help in treating critically sick patients with COVID-19.

2.5 HAEMATOLOGICAL ABNORMALITIES

Thrombotic events are more likely to occur in individuals who are already critically sick if they get COVID-19 during the acute phase of the disease. [26] Hypoxia-inducible transcription factor overexpression, microvascular dysfunction and enhanced expression of tissue factors in response to inflammatory cytokines all play a role in the development of this coagulopathy. [27–29] Patients with COVID-19 are at a greater risk of thrombosis, therefore the study compared the routine use of intermediate-dose anticoagulation (enoxaparin 1 mg/kg daily) to conventional prophylactic dosage anticoagulation (enoxaparin 40 mg daily) in critically sick patients. Comparing intermediate-dose anticoagulation to conventional prophylactic-dose anticoagulation did not indicate a reduction in the composite outcome of arterial or venous thrombosis, ECMO use or death after 30 days. [30] Despite the presence of bleeding episodes, the advantages of in-hospital VTE prevention outweigh the dangers for most patients. [31]

It is unclear how long hypercoagulability lasts, although most cases of VTE appear to occur within the first two to four weeks after infection. It has been proposed to provide VTE prophylaxis to COVID-19 patients in the outpatient environment after they are released from the hospital; however, this is not supported by current recommendations. Studies have demonstrated a decreased incidence of VTE, but an increased risk of bleeding, making the overall benefit of longer duration pharmacological thromboprophylaxis in severely unwell medical patients uncertain. Prophylaxis would be beneficial in patients with COVID-19 who are at low risk of bleeding if the rate of symptomatic VTE at 35 to 42 days post-discharge was more than 1.8%, [32] but current data does not suggest that this is the case.

Rates of VTE in hospitalised patients with COVID-19 varied from 0.48% percent to 1.9% percent in retrospective investigations. These rates may be comparable to those seen in other hospital-discharged patient populations with symptomatic VTE. Risks of VTE, arterial thromboembolism and severe bleeding were 1.55%, 1.71% and 1.73%, respectively, in a large prospective research aimed at better assessing various post-discharge haematological outcomes in patients with COVID-19. Although these authors found that patients with COVID-19 who received pharmacological thrombo-prophylaxis after discharge had a lower risk of death, venous thromboembolism (VTE) and arterial thromboembolism (ATE), they did not evaluate whether this treatment significantly increased the risk of bleeding.

2.6 NEUROLOGICAL COMPLICATIONS

Long-term effects following SARS-CoV-2 infection have been documented in the fields of neuroscience and mental health. Multiple sources of data on long-term symptoms in individuals two months after acute infection showed persistent neurological abnormalities, such as tiredness, muscular weakness, sleep problems, myalgia and headache. [33] These symptoms are now typically associated with the long COVID condition. SARS-CoV-2 infection is distinctive from other viral infections in that it causes a loss of smell and taste. COVID-19 is related with a high rate of severe, critical illness and acute respiratory distress syndrome (ARDS), which is associated with cognitive disturbances comparable to those shown in individuals with ARDS in previous research. At one year of follow-up, ARDS survivors from other sources showed impairments in memory (13%), verbal fluency (16%) and executive function (49%).

People infected with SARS-CoV-2 also tend to suffer from long-term psychological effects. At the 60-day follow-up after hospitalisation for COVID-19, 23% of patients were still experiencing anxiety or sadness, according to a study by Huang et al. Post-traumatic stress disorder (PTSD) was found to affect 28% of the 402 individuals studied by Mazza et al., [34] with depression affecting 31%, anxiety affecting 42% and sleeplessness affecting 40%. The incidence of a new mental diagnosis in the 14–90 days following SARS-CoV-2 infection was higher than that of various other health events (such as other respiratory infections, skin infection, bone fracture, etc., according to a large-scale cohort research conducted in the United States. [35]

Similar to the disease reported in other organs, it is believed that many pathways contribute to COVID-19-related neurological problems. Direct viral damage, systemic inflammation and cerebrovascular alterations are only some of the putative processes, with some combination of these factors being the most plausible explanation. [36] However, to yet, no autopsy investigations validating this conclusion in SARS-CoV-2 have been reported. Indeed, SARS-CoV-2 was not detected in cerebral fluids tested by RT-PCR from severely sick individuals. [37] These results provide credence to the theory that neurocognitive problems following SARS-CoV-2 infection are caused in large part by systemic inflammatory damage. Pathophysiology of taste and smell impairment is poorly understood. Similarly, it has been hypothesised that widespread ACE2 expression on tongue and other oral mucous membranes contributes to taste loss. Due to the close relationship between smell and taste, it has been postulated that olfactory impairment greatly contributes to taste impairment. [38]

Any patient exhibiting neurological or psychiatric symptoms after SARS-CoV-2 infection should be further evaluated with thorough neuropsychological testing. Screening for anxiety and depression using standard instruments and conducting additional PTSD screenings in high-risk populations (those with severe illness) may help explain some of the observed cognitive changes. Patients with a poor sense of smell or taste may benefit from olfactory training, which entails daily exposure to strong odours for three months.

2.7 DERMATOLOGICAL SEQUELAE

A literature study indicated that maculopapular exanthem (morbilliform) was the most common cutaneous manifestation of COVID-19 infection, reported by

36.1% of 72 documented individuals across 18 studies, followed by papulovesicular rash (34.7%), urticaria (9.7%) and painful red acral purple papules (15.3%), with 19.4% of these manifestations being in the hands and feet (67). Pernio-like lesions were shown to be the most common cutaneous manifestation in another multinational investigation of 2,560 patients (51.5%), with a latency interval of 1.5 days in children and 7.9 days in adults between upper respiratory infections and cutaneous findings. [39] Hair loss was considerably more commonly observed for patients months after COVID-19 infection, with 24 of 120 patients (20.0%) reporting it as a post-discharge symptom 110 days after hospital discharge in the Chinese post-acute COVID-19 study of hospitalised patients. In spite of this, case reports have documented unusual presentations, implying that even among people infected with the same virus, there may be variations in how it manifests itself. [40]

Many of these symptoms are likely caused by the same processes as other long-haul COVID-19 symptoms, but there are some distinct viral pathways that may be relevant. For example, telogen effluvium is the medical term for the transient hair loss caused by nonscarring alopecia that can occur after a stressful event, hormonal changes after childbirth or an acute febrile illness/viral infection that lasts less than six months. With the use of drugs like minoxidil, finasteride and topical corticosteroids, patients who have had hair loss due to a COVID-19 infection may discover that their symptoms improve. [41]

2.8 DIABETES-RELATED COMORBIDITIES

The presence of pre-existing diabetes mellitus has been linked to poorer COVID-19 outcomes. Patients with both type 1 and type 2 diabetes have shown an increased risk of developing new-onset hyperglycemia and acute decompensation of diabetes, including diabetic ketoacidosis, after exposure to COVID-19. Insulin resistance as a result of the inflammatory state and insulin secretory deficits from damaged cells (due to direct virus damage or indirect effects) is postulated reasons for hyperglycemia after infection in addition to iatrogenic hyperglycemia from steroid usage. [42] It is unknown what percentage of those who were diagnosed with diabetes after COVID-19 had the disease before infection but had it unmasked or made worse. New-onset diabetes after hospitalisation for COVID-19 is not known to be a persistent condition. As a result, the worldwide CoviDiab Registry was established to further quantify the length of time post-COVID-19 diabetes lasts and to expand upon the findings of previous studies on the correlation between the two conditions.

2.9 RENAL COMPLICATIONS

Patients with acute COVID-19 often suffer from acute kidney injury (AKI), necessitating in-hospital dialysis or a transplant. [43] Direct viral damage, systemic hypoxia, the impact of inflammatory cytokines and aberrant coagulation are all contributors to the aetiology of acute kidney injury. [44] Histopathologically, acute tubular necrosis is the norm, albeit glomerulopathy and microvascular thrombi are also detected. [45] Hospital mortality rises in patients with AKI, and even those who recover may be left with kidney damage. One study found that even after being released from the hospital, 35% of patients with AKI still had impaired renal function, and 30% of those who had needed in-hospital dialysis continued to do so at home. Remarkably, 36% of individuals with residual kidney illness at discharge had recovered by the time of the median 21-day follow-up in the same trial, while 14% of those who recovered before discharge had recurring kidney disease. [46] Another study found that at six months, 35% of COVID-19 survivors had renal impairment, and 13% had new-onset renal dysfunction after having normal kidney function during their initial illness. Acute COVID-19 renal disease patients should begin seeing a nephrologist for therapy because of the positive correlation between early detection and better results. [47]

2.10 GASTROINTESTINAL SEQUELAE

Acute COVID-19 often causes and is accompanied by gastrointestinal distress. Diarrhoea was reported by 6% of those with post-acute COVID-19 symptoms, making it one of the top ten most prevalent complaints. Other chronic symptoms include throwing up, feeling sick, having stomach pain and losing your appetite. [48] The prolonged faecal shedding of SARS-CoV-2 observed even after respiratory samples became negative [49] suggests that continued virus replication in the gastrointestinal system may be responsible for these lingering symptoms. Alterations in the gut flora are also linked to COVID-19, but the impact on long-term symptoms is unclear. [50]

Hepatocellular damage and/or biliary stasis, which manifest as abnormal liver function tests in the acute phase, are common. Direct viral cytotoxicity, especially in the biliary tree, as well as systemic inflammation, hypoxia, coagulopathy and unfavourable effects of medicines, are among the mechanisms of harm. Hepatic steatosis, hepatic congestion, arterial thrombosis, fibrosis, portal inflammation and lobular inflammation are all often observed in autopsy studies. [51] Abnormalities in liver function may continue in survivors of COVID-19 with acute liver injury, but they often improve over the course of weeks to months.

2.11 CONCLUSION

Regarding the COVID-19 pandemic, there are still many open questions. Novel viral variants, such as the Delta (B.1.617.2 lineage) variant first detected in India, Gamma (P. 1 lineage) first identified in Brazil and Lambda (C.37 lineage) first identified in Peru, have emerged due to the lack of effective measures to contain the pandemic (2). The World Health Organization has identified a number of these novel variants as variants of interest (VOIs), which may indicate that treatments and vaccinations are less effective against them. Notably, a number of breakthrough cases of the Delta variant have been found even among vaccinated individuals in the United States. This is likely due to the fact that the Delta variant is able to partially, but not completely, avoid the neutralising monoclonal or polyclonal antibodies elicited by immunity to SARS-CoV-2, either through vaccination or previous infection.

Given the recent emergence of these novel variations, it is yet unknown how long-haul COVID-19 infection will manifest in infected individuals. Because the Delta variety is so new, there have been no extensive published studies comparing how patients initially appear with the condition and how they fare over time. According to a study of the Delta variant in China published as a preprint, 73.9% of transmissions from Delta cases occur before symptom manifestation. Nonetheless, vaccination appeared to be effective in this population as well, as those who were either unvaccinated (OR: 2.84) or who had only had one dose of vaccination (OR: 6.02) were more likely to spread infection than those who had received two doses of immunisation. More extensive research are needed to determine if or not there are differences in the symptoms experienced by patients of different ages and races who have these new variations.

Finally, prior research on the Alpha form of the virus indicated ethnic differences in treatment response, symptom duration and risk of getting COVID-19, in addition to prolonged COVID-19 symptoms. It is not known if these variations are the result of different approaches to treatment or whether there are underlying molecular pathways that predispose some people to experience more persistent symptoms. Further research is needed to determine the effects of these novel strains on the progression of protracted COVID-19 and the kind of patients that are most at risk.

REFERENCES

1. Dong E, Du H, Gardner L. An interactive web-based dashboard to track COVID-19 in real time. Lancet Infect Dis 20: 533–534, 2020.
2. Faria NR, Mellan TA, Whittaker C et al. Genomics and epidemiology of a novel SARS-CoV-2 lineage in Manaus, Brazil. medRxiv, 2021.
3. Planas D, Veyer D, Baidaliuk A et al. Reduced sensitivity of SARS-CoV-2 variant delta to antibody neutralization. Nature 596: 276–280, 2021.
4. Centers for Disease Control and Prevention. COVID-19 Weekly Cases and Deaths per 100,000 Population by Age, Race/Ethnicity, and Sex. 2021.
5. Nalbandian A, Sehgal K, Gupta A et al. Post-acute COVID-19 syndrome. Nat Med 27: 601–615, 2021.
6. SeyedAlinaghi S, Afsahi AM, MohsseniPour M et al. Late complications of COVID-19; a systematic review of current evidence. Arch Acad Emerg Med 9: e14, 2021.
7. Sudre CH, Murray B, Varsavsky T et al. Attributes and predictors of long COVID. Nat Med 27: 626–631, 2021.
8. Gupta A, Madhavan MV, Sehgal K et al. Extrapulmonary manifestations of COVID-19. Nat Med 26: 1017–1032, 2020.
9. Hoffmann M, Kleine-Weber H, Schroeder S et al. SARS-CoV-2 cell entry depends on ACE2 and TMPRSS2 and is blocked by a clinically proven protease inhibitor. Cell 181: 271–280, 2020.

10. Qi F, Qian S, Zhang S et al. Single cell RNA sequencing of 13 human tissues identify cell types and receptors of human coronaviruses. Biochem Biophys Res Commun 526: 135–140, 2020.
11. Pan X-W, Xu D, Zhang H et al. Identification of a potential mechanism of acute kidney injury during the COVID-19 outbreak: A study based on single-cell transcriptome analysis. Intensive Care Med 46: 1114–1116, 2020.
12. Ziegler CGK, Allon SJ, Nyquist SK et al. SARS-CoV-2 receptor ACE2 is an interferon-stimulated gene in human airway epithelial cells and is detected in specific cell subsets across tissues. Cell 181: 1016–1035, 2020.
13. Jin Y, Ji W, Yang H et al. Endothelial activation and dysfunction in COVID-19: From basic mechanisms to potential therapeutic approaches. Signal Transduct Target Ther 5: 293, 2020.
14. Zhao Y-M, Shang Y-M, Song W-B et al. Follow-up study of the pulmonary function and related physiological characteristics of COVID-19 survivors three months after recovery. EClinicalMedicine 25: 100463, 2020.
15. Ehrenfeld M, Tincani A, Andreoli L et al. Covid-19 and autoimmunity. Autoimmun Rev 19: 102597, 2020.
16. Ngai JC, Ko FW, Ng SS et al. The long-term impact of severe acute respiratory syndrome on pulmonary function, exercise capacity and health status. Respirology 15: 543–550, 2010.
17. Shang J, Ye G, Shi K et al. Structural basis of receptor recognition by SARS-CoV-2. Nature 581: 221–224, 2020.
18. Carfì A, Bernabei R, Landi F et al. Against COVID-19 post-acute care study group. Persistent symptoms in patients after acute COVID-19. JAMA 324: 603–605, 2020.
19. Raj SR, Arnold AC, Barboi A et al. Long-COVID postural tachycardia syndrome: An American autonomic society statement. Clin Auton Res 31: 365–368, 2021.
20. Puntmann VO, Carerj ML, Wieters I et al. Outcomes of cardiovascular magnetic resonance imaging in patients recently recovered from coronavirus disease 2019 (COVID-19). JAMA Cardiol 5: 1265–1273, 2020.
21. Kang Y, Chen T, Mui D et al. Cardiovascular manifestations and treatment considerations in COVID-19. Heart 106: 1132–1141, 2020.
22. Zou X, Chen K, Zou J et al. Single-cell RNA-seq data analysis on the receptor ACE2 expression reveals the potential risk of different human organs vulnerable to 2019-nCoV infection. Front Med 14: 185–192, 2020.
23. Siripanthong B, Nazarian S, Muser D et al. Recognizing COVID-19-related myocarditis: The possible pathophysiology and proposed guideline for diagnosis and management. Heart Rhythm 17: 1463–1471, 2020.
24. Bozkurt B, Kovacs R, Harrington B. Joint HFSA/ACC/AHA statement addresses concerns Re: Using RAAS antagonists in COVID-19. J Card Fail 26: 370, 2020.
25. Helms J, Tacquard C, Severac F et al. High risk of thrombosis in patients with severe SARS-CoV-2 infection: A multicenter prospective cohort study. Intensive Care Med 46: 1089–1098, 2020.
26. Klok FA, Kruip MJHA, van der Meer NJM et al. Confirmation of the high cumulative incidence of thrombotic complications in critically ill ICU patients with COVID-19: An updated analysis. Thromb Res 191: 148–150, 2020.
27. Hadid T, Kafri Z, Al-Katib A. Coagulation and anticoagulation in COVID-19. Blood Rev 47: 100761, 2021.
28. Lazzaroni MG, Piantoni S, Masneri S et al. Coagulation dysfunction in COVID-19: The interplay between inflammation, viral infection and the coagulation system. Blood Rev 46: 100745, 2021.

29. Bikdeli B. Intermediate versus standard-dose prophylactic anticoagulation in critically ill patients with COVID-19—INSPIRATION-S. In: American College of Cardiology Virtual Annual Scientific Session (ACC 2021).
30. Paranjpe I, Fuster V, Lala A et al. Association of treatment dose anticoagulation with in-hospital survival among hospitalized patients with COVID-19. J Am Coll Cardiol 76: 122–124, 2020.
31. Al-Samkari H, Karp Leaf RS, Dzik WH et al. COVID-19 and coagulation: Bleeding and thrombotic manifestations of SARS-CoV-2 infection. Blood 136: 489–500, 2020.
32. Moores LK, Tritschler T, Brosnahan S et al. Prevention, diagnosis, and treatment of VTE in patients with coronavirus disease 2019: CHEST guideline and expert panel report. Chest 158: 1143–1163, 2020.
33. Huang C, Huang L, Wang Y et al. 6-month consequences of COVID-19 in patients discharged from hospital: A cohort study. Lancet 397: 220–232, 2021.
34. Mazza MG, De Lorenzo R, Conte C et al. COVID-19 BioB outpatient clinic study group. Anxiety and depression in COVID-19 survivors: Role of inflammatory and clinical predictors. Brain Behav Immun 89: 594–600, 2020.
35. Taquet M, Luciano S, Geddes JR et al. Bidirectional associations between COVID-19 and psychiatric disorder: Retrospective cohort studies of 62 354 COVID-19 cases in the USA. Lancet Psychiat 8: 130–140, 2021.
36. Heneka MT, Golenbock D, Latz E et al. Immediate and long-term consequences of COVID-19 infections for the development of neurological disease. Alzheimers Res Ther 12: 69, 2020.
37. Helms J, Kremer S, Merdji H et al. Neurologic features in severe SARS-CoV-2 infection. N Engl J Med 382: 2268–2270, 2020.
38. Vaira LA, Salzano G, Fois AG et al. Potential pathogenesis of ageusia and anosmia in COVID-19 patients. Int Forum Allergy Rhinol 10: 1103–1104, 2020.
39. Mirza FN, Malik AA, Omer SB, Sethi A. Dermatologic manifestations of COVID-19: A comprehensive systematic review. Int J Dermatol 60: 418–450, 2021.
40. Killion L, Beatty PE, Salim A. Rare cutaneous manifestation of COVID-19. BMJ Case Rep 14: e240863, 2021.
41. Harrison S, Sinclair R. Telogen effluvium. Clin Exp Dermatol 27: 389–385, 2002.
42. Unnikrishnan R, Misra A. Diabetes and COVID19: A bidirectional relationship. Nutr Diabetes 11: 21, 2021.
43. Robbins-Juarez SY, Qian L, King KL et al. Outcomes for patients with COVID-19 and acute kidney injury: A systematic review and meta-analysis. Kidney Int Rep 5: 1149–1160, 2020.
44. Su H, Yang M, Wan C, Yi L-X et al. Renal histopathological analysis of 26 postmortem findings of patients with COVID-19 in China. Kidney Int 98: 219–227, 2020.
45. Santoriello D, Khairallah P, Bomback AS et al. Postmortem kidney pathology findings in patients with COVID-19. J Am Soc Nephrol 31: 2158–2167, 2020.
46. Chan L, Chaudhary K, Saha A et al. COVID informatics center (MSCIC). AKI in hospitalized patients with COVID-19. J Am Soc Nephrol 32: 151–160, 2021.
47. Meier P, Bonfils RM, Vogt B et al. Referral patterns and outcomes in noncritically ill patients with hospital-acquired acute kidney injury. Clin J Am Soc Nephrol 6: 2215–2225, 2011.
48. Aiyegbusi OL, Hughes SE, Turner G et al. TLC study group. Symptoms, complications and management of long COVID: A review. J R Soc Med 114: 428–442, 2021.
49. Xu Y, Li X, Zhu B et al. Characteristics of pediatric SARS-CoV-2 infection and potential evidence for persistent fecal viral shedding. Nat Med 26: 502–505, 2020.

50. Spearman CW, Aghemo A, Valenti L, Sonderup MW et al. COVID-19 and the liver: A 2021 update. Liver Int 41: 1988–1998, 2021.
51. Díaz LA, Idalsoaga F, Cannistra M et al. High prevalence of hepatic steatosis and vascular thrombosis in COVID-19: A systematic review and meta-analysis of autopsy data. World J Gastroenterol 26: 7693–7706, 2020.

Part II

SARS CoV-2 Co-Infection with Other Pathogens

3 Falciparum Malaria and COVID-19 Co-Infections

Aashna Srivastava, Tabish Qidwai and Mariam Imam

3.1 INTRODUCTION

Since December 2019, the COVID-19 pandemic has expanded from its epicentre in Wuhan, China, infecting over 3 million individuals and killing over 200,000 people globally. SARS-CoV-2 is the pathogen that caused the disease. It is an enveloped RNA beta coronavirus that is phylogenetically linked to SARS-CoV-1. Fever and cough are the most typical symptoms; more severe results (needing intensive care and mechanical breathing) are linked to older age, a higher percentage of comorbidities and higher mortality. [1] The percentage of SARS-CoV-2 infected people who stay asymptomatic throughout the course of infection has not yet been conclusively determined. Clinical indications of the illness, which include fever, coughing, nasal congestion, exhaustion and other symptoms of upper respiratory tract infections, typically begin less than a week after the onset of symptoms in symptomatic patients. Approximately 75% of patients may develop a severe illness with dyspnea and severe chest symptoms similar to pneumonia as a result of the infection, as determined by computed tomography at the time of admission. The second or third week after a symptomatic infection is when pneumonia typically develops. The most prominent signs of viral pneumonia include decreased oxygen saturation, blood gas deviations and alterations seen on chest X-rays and other imaging modalities, such as ground glass anomalies, patchy consolidation, alveolar exudates and interlobular involvement, which finally indicate worsening. Inflammatory indicators including C-reactive protein and proinflammatory cytokines are high, and lymphopenia seems to be a common condition. [2]

Malaria is an infectious disease caused by Plasmodium protozoa, which can be fatal if not diagnosed and treated promptly. Malaria is a mosquito-borne parasite disease caused by Plasmodium falciparum that affects 200 to 400 million people worldwide each year, resulting in almost 400,000 fatalities each year and disproportionately affecting children in Sub-Saharan Africa. Although public health measures such as insecticide-treated bed nets and anti-malarial medications helped to reduce the number of cases of malaria by 50 to 75% globally between 2000 and 2015, the incidence of the disease is now rising in many regions in spite of these interventions. [3]

DOI: 10.1201/9781003324911-5

COVID-19 is a newly emerging disease that can affect different body systems; however, the respiratory system is mainly reported. [4] The COVID-19 outbreak serves as a stark reminder of the ongoing challenge posed by emerging and re-emerging infectious pathogens and the need for ongoing monitoring, quick diagnosis and in-depth research in order to comprehend the basic biology of novel organisms, our susceptibilities to them and to develop efficient countermeasures. [5] The SARS-CoV-2 virus primarily affects the respiratory system, although it also affects other organ systems. In the original case series, signs of lower respiratory tract infections such as fever, dry cough and dyspnea were noted. In addition, many reported experiencing headaches, vertigo, widespread weakness, vomiting and diarrhoea. It is now widely recognised that COVID-19's respiratory symptoms can range from hardly perceptible to severely hypoxic with ARDS. [6] COVID-19 or malaria can cause severe disease, and both are highly infectious; then, co-infection is expected in malaria-endemic regions, and it's expected to be even more fatal than either of the two COVID-19 or malaria isolated.

This chapter aims to establish the prevalence of malaria and describe the clinical characteristics of SARS-CoV-2 and *P. falciparum* co-infection in a high burden malaria setting. Further, we aimed to cover the information on the prevalence and characteristics of malaria infection among COVID-19-infected individuals. The findings will help us better understand this particular comorbidity during the COVID-19 pandemic.

3.2 INTERACTIONS BETWEEN SARS-CoV-2 INFECTION AND MALARIA

As COVID-19 cases continue to rise globally, there is increasing concern that a more severe COVID-19 course could be seen among malaria-infected individuals. However, the clinical and immunological responses of malaria-infected patients to COVID-19 are yet to be specifically investigated. The potential for the disruptive effect of COVID-19 on health systems which are most fragile in malaria-endemic countries, potentially undermining malaria treatment and prevention effort in these countries, is also of concern. In addition, malaria and COVID-19 have some common symptoms, such as fever, headache, body weakness and aches. Therefore, malaria endemicity may complicate clinical diagnosis of COVID-19 especially in areas where access to testing is insufficient. Hemoglobin, RBC count and other RBC-related indicators typically fall in cases of severe malaria. These characteristics, together with the patient's symptoms, suggest to co-infection as a possible cause of hepatic encephalopathy. However, because a patient's blood contains malaria parasites, the neurological consequences in the patient could be a result of cerebral malaria. SARS-CoV-2 has the ability to change how the body's endothelial cells operate, according to reports. Endothelial activation can happen during the COVID-driven cytokine storm or as a result of viral infection of the endothelial cells themselves. Similar endothelial activation and increased surface adhesion molecules are seen in malaria cases. [7]

COVID-19 and malaria patients have two times as many pathogens burdening them. In severe cases, COVID-19 has an impact on immunological cells, causing a cytokine storm, which further activates endothelial cells. RBCs infected with Plasmodium have the ability to adhere to endothelial cells and activate them as well. Endothelial cells may express more cell surface receptors as a result of an accentuated host response. In order to sequester infected RBCs, an overwhelming number of Plasmodium-infected and uninfected RBCs eventually cling to the endothelium wall. Additionally, endothelial cells can influence the activity of neighbouring astrocytes, which are activated in response to injury at the blood-brain barrier. Increased barrier permeability, leukocyte extravasation, and severe localised inflammation may be made possible by this. [7]

3.3 SIMILARITY OF MALARIA AND COVID-19 SYMPTOMS AND PROBLEMS ASSOCIATED WITH IT

The symptoms of malaria and COVID-19 are similar, including fever, chills, headache, sweating, vomiting and body throbs. When a patient displays this combination of symptoms, it raises questions for medical professionals trying to diagnose the illness. Malaria and COVID-19 co-infection could lead to excessive proinflammatory response, leading to severe manifestations and poor prognosis. COVID-19 infection and malaria share a number of clinical symptoms, particularly extrapulmonary symptoms such thrombocytopenia, raised transaminase enzymes, increased bilirubin and impaired kidney function. The symptoms of hepatic problems in COVID-19 infection are brought on by a direct viral infection of the liver cells, which is shown by elevated levels of aspartate aminotransferase and/or alanine aminotransferase. About 19–53% of COVID-19 individuals showed this increment. Additionally, a study indicated that one patient's alkaline

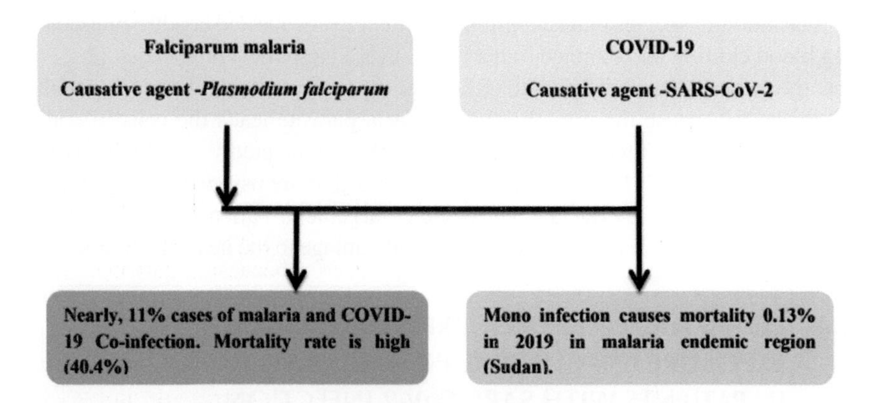

FIGURE 3.1 A co-infection of COVID-19 and malaria causes high mortality in endemic regions.

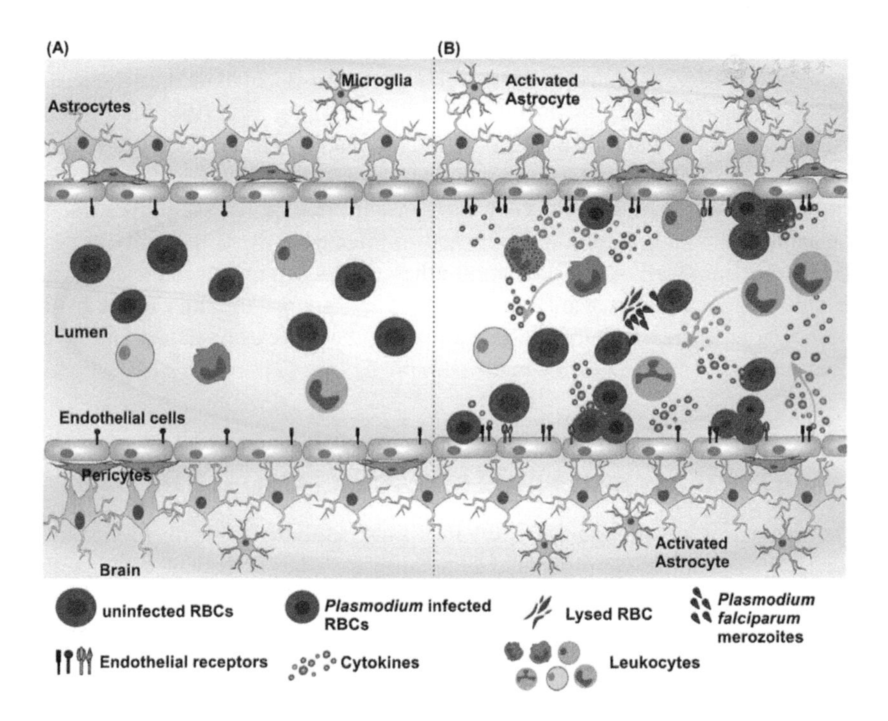

FIGURE 3.2 Representation of possible microenvironment at blood-brain barrier in (A) normal individual, (B) Malaria and COVID-19 positive individual. Individual suffering from COVID-19 and malaria experience double pathogen burden. [7]

phosphatase level increased, and 30 out of 56 patients exhibited an increase in gammaglutamyl transferase. Patients with COVID-19 have a bad prognosis because of their elevated bilirubin levels and jaundice. [8]

Plasmodium species that cause malaria infect liver cells and erythrocytes, causing blood clotting and damage to the heart, liver and kidney. It has been observed that the red cell distribution width (RDW) of erythrocytes increases during malarial invasion because the growth of the malaria parasite leads the cells to widen. Additionally, it was found that increasing RDW is associated with a higher risk of mortality in COVID-19. There are other factors that are associated with malaria infection than the COVID-19 erythrocyte enlargement. Patients with COVID-19 also showed symptoms like blood clotting and damage to the heart, liver or kidney.

3.4 POTENTIAL EFFECTS OF PREVIOUS MALARIA EXPOSURE ON CLINICAL PROFILES AND OUTCOMES IN PATIENTS WITH SARS-CoV-2 INFECTION

Results have revealed statistically significant elevation of 8-iso PGF2α in COVID-19 patients coinfected with malaria compared to COVID-19 patients only, and this may be due to increase production of free radicals. Furthermore, a significant decrease of alpha-tocopherol was observed in COVID-19 coinfected with malaria compared to COVID-19 patients due to increase utilisation of anti-oxidants in counterbalancing the negative effect of free radicals generated.

The association between malaria and COVID-19 has led to speculation that the low COVID-19 cases in countries where malaria is endemic may be attributable to the anti-malaria immunity, which offers variable protection against SARS-CoV-2. The apparent immunodominant T cell epitope conservation between the conserved surface protein of P. malariae and the SARS-CoV-2 spike glycoprotein may confer immunity on those who have already contracted Plasmodium. [9]

3.5 IMPORTANT CLINICAL CORRELATIONS OF CO-INFECTION IN HIGH MALARIA BURDEN AREAS

It has been suggested that the anti-malaria medication chloroquine (CQ) and its derivative hydroxychloroquine (HCQ) could be used to treat COVID-19. Both drugs have demonstrated some effectiveness in decreasing viral replication in SARS-CoV and MERS-CoV, both in vitro and in pre-clinical investigations, with HCQ being more effective than CQ in inhibiting SARS-CoV-2 in vitro. Although in vivo, additional studies are needed before the medications may be approved as COVID-19 therapies or preventives. Both CQ and HCQ medications have been shown to be effective against malaria, assisting programmes to control and eradicate the disease. Chloroquine resistance eventually developed as a result of the widespread usage of CQ, initially along the Cambodia-Thailand border in the 1950s to 1960s, and then it was recorded all over the world. Despite this problem, both medications are nevertheless employed as preventive and therapeutic measures in the majority of P. vivax endemic areas. [10]

3.6 CONCLUSION

This broad study highlights the need for identifying possible underlying secondary infections in conjunction with SARS-CoV-2, which may otherwise go unnoticed in the midst of the present extraordinary COVID-19 pandemic. It is critical for doctors to keep in mind that neither the presence of COVID-19 co-infection nor the presence of other infections cannot be totally ruled out when COVID-19 is confirmed. Co-infections in COVID-19 are relatively unknown, with newly emerging evidence suggesting simultaneous infections with influenza, varicella-zoster and respiratory syncytial virus (RSV). In order to detect signs of SARS-CoV-2 co-infection with malaria or other NTDs, it will be crucial to create surveillance platforms quickly. The current healthcare system must be strengthened to accommodate an influx of several sick patients, particularly by putting measures in place to provide care for foreign workers and immigrants regardless of their legal status or papers. Moreover, additional research is required in order to fully comprehend the pathophysiology and potential clinical significance of this co-infection.

REFERENCES

1. Lansbury L, Lim B, Baskaran V, Lim WS. Co-infections in people with COVID-19: A systematic review and meta-analysis. *Journal of Infection*. 2020 Aug 1;81(2):266–275.
2. Velavan TP, Meyer CG. The COVID-19 epidemic. *Tropical Medicine & International Health*. 2020 Mar;25(3):278.
3. Gaudinski MR, Berkowitz NM, Idris AH, Coates EE, Holman LA, Mendoza F, Gordon IJ, Plummer SH, Trofymenko O, Hu Z, Campos CA. A monoclonal antibody for malaria prevention. *New England Journal of Medicine*. 2021 Aug 26;385(9):803–814.
4. World Health Organization. Coronavirus disease (COVID-19) pandemic 2021 [cited April 6, 2021]. Available from: www.who.int/emergencies/diseases/novel-coronavirus-2019.
5. Fauci AS, Lane HC, Redfield RR. Covid-19—navigating the uncharted. *New England Journal of Medicine*. 2020;382(13):1268–1269.
6. Yuki K, Fujiogi M, Koutsogiannaki S. COVID-19 pathophysiology: A review. *Clinical Immunology*. 2020;215:108427.
7. Indari O, Baral B, Muduli K, Mohanty AP, Swain N, Mohakud NK, Jha HC. Insights into Plasmodium and SARS-CoV-2 co-infection driven neurological manifestations. *Biosafety and Health*. 2021 Aug 25;3(4):230–234.
8. Katu S, Ilyas M, Daud N. Case report: Covid-19 and severe malaria co-infection. *European Journal of Molecular & Clinical Medicine*. 2020 Dec 2;7(8):961–968.
9. Hassan MM, Sharmin S, Hong J, Lee HS, Kim HJ, Hong ST. T cell epitopes of SARS-CoV-2 spike protein and conserved surface protein of Plasmodium malariae share sequence homology. *Open Life Sciences*. 2021 Jan 1;16(1):630–640.
10. Zawawi A, Alghanmi M, Alsaady I, Gattan H, Zakai H, Couper K. The impact of COVID-19 pandemic on malaria elimination. *Parasite Epidemiology and Control*. 2020 Nov 1;11:e00187.

4 Interaction of COVID-19 and Tuberculosis

Malti Dadheech and Anand Kumar Maurya

4.1 INTRODUCTION

Tuberculosis (TB) is the oldest disease still affecting humanity and is one of the leading causes of morbidity and mortality, worldwide (Alene et al., 2020). TB was the major cause of mortality from any infectious agent till the emergence of the COVID-19. Approximately, a quarter of the global population has a latent infection of TB (Kyu et al., 2018). According to the World Health Organisation (WHO) TB report, 2021, an estimated 9.9 million people fell ill with TB, with an estimated 1.3 million deaths among HIV-negative cases and 214,000 deaths among HIV-positive cases in the year 2020, globally (Global Tuberculosis Report 2021, n.d.)

In December 2019, the novel coronavirus was identified as the cause of a cluster of pneumonia cases in Wuhan, a city in the Hubei Province of China. In February 2020, the World Health Organization designated the disease COVID-19, which stands for coronavirus disease 2019 (World Health Organization, 2020). COVID-19 is caused by SARS-CoV-2, a member of the beta Coronaviridae family, which also includes SARS-CoV-1 (severe acute respiratory syndrome coronavirus 1) and MERS-CoV (Middle East respiratory syndrome coronavirus) (Lu et al., 2020b). Over 3 billion individuals, or 44% of the global population, had been infected with COVID-19 at least once (McIntosh et al., 2021). The emergence of COVID-19 disease considered a pandemic by WHO has created a threat to humanity and has created new challenges for global as well as national TB control efforts. The TB/SARS-CoV2 co-infection has also raised concern for health care authorities, especially in high TB endemic countries. The WHO set the goal, including a 90% reduction in TB incidence and a 95% reduction in TB deaths compared with 2015 and no catastrophic costs due to TB by 2035 (Uplekar et al., 2015).

In order to achieve the targets placed by the WHO, the COVID-19 pandemic could be a major challenge. Considering the disruption caused by COVID-19 globally, it is crucial to consider the possible effects of controlling and preventing TB, which might be even more disastrous to public health than COVID-19. The impact of the COVID-19 disease on diseases such as diabetes and cancer has been addressed, but there are no review articles on the impact of COVID-19 on TB (Hanna et al., 2020; Maddaloni & Buzzetti, 2020).

DOI: 10.1201/9781003324911-6

The early symptoms of TB and COVID-19 are very similar, which can interrupt the diagnosis and treatment of both diseases (Prasad et al., 2020). A study described the timing of diagnosis of the duo, observing that one-third had COVID-19 diagnosed before TB and 18% were diagnosed simultaneously (Tadolini, García-García, et al., 2020). The COVID-19 pandemic has reversed decades of progress in decreasing the number of people who die from TB infection. The home isolation of COVID-19 infected patients could promote TB transmission in low-socioeconomic and overcrowded conditions. For that reason, considering the potential impacts of COVID-19 on TB is important for designing preventive approaches (Alene et al., 2020). According to a study, patients with TB and its co-infection with SARS-CoV2 are also associated with a high rate of mortality (Song et al., 2021).

M. tuberculosis and its coinfection with SARS-CoV2 are associated with diagnostic challenges in terms of poor performance (lower sensitivity) for smear microscopy, frequent extra-pulmonary localisation and atypical chest radiography appearance. Conventional methods of TB diagnosis such as culture are the gold standard but are relatively slow and take almost six to eight weeks (Parsons et al., 2011).

The overlapping clinical manifestations of this duo could also lead to a delayed or missed diagnosis (Aznar et al., 2021). However, modern molecular methods are relatively rapid, more sensitive and helpful for the early diagnosis and initiation of treatment therapy. Therefore, easy-to-use molecular techniques are urgently required in clinical settings in high burden as well as remote areas and resource-constrained settings. This chapter mainly focuses on the diagnostic tools, biological mechanism and interaction of TB with COVID-19 disease.

4.2 EPIDEMIOLOGICAL AND CLINICAL PRESENTATION OF COVID-19 WITH TB INFECTION

TB and COVID-19 both pose a significant burden with the rate of cases worldwide. In an interesting meta-analysis study from China, the prevalence of TB in COVID-19-infected patients ranged from 0.47 to 4.47% (Gao et al., 2021). The prevalence of TB disease was significantly increased in cases with severe COVID-19 than in non-severe cases. In a cohort study from eight countries (Tadolini, Codecasa, et al., 2020), it was concluded that patients can be infected with COVID-19 before or after the TB disease diagnosis. The role of COVID-19 in increasing active TB development and vice versa is not yet clear. More studies need to be done to understand the evolution and outcomes in patients with both diseases. The above-mentioned findings were confirmed by a similar study conducted in India (Gupta et al., 2020). The reference TB centre in Italy conducted a clinical study (Stochino et al., 2020) in which it was concluded that the effect of COVID-19 on active TB cases can be manageable with proper treatment and patient care. Control of the infection and medical devices is very important to stop the risk in the hospitals. In most of the young-age patients, clinical and radiological deterioration may be associated (Ippolito et al., 2020).

4.3 BIOLOGICAL MECHANISM OF COVID-19 AND TB

The genome of the SARS-CoV-2 virus is similar up to 80% to SARS-CoV-1 and 50% to MERS-CoV (Gorbalenya, A. E et al., 2020). The S protein belongs to the class I viral fusion proteins, which over-expresses the angiotensin-converting enzyme 2 (ACE2) as the receptor entering into the human host (Hoffmann et al., 2020; Millet & Whittaker, 2014). COVID-19 can be loosely split into three stages: 1) asymptomatic incubation period with or without detectable virus; 2) non-severe symptomatic period with the presence of virus; 3) severe respiratory symptomatic stage with high viral load (D. Wang et al., 2020) and subsequent deterioration of the lung damage, respiratory failure (that may require invasive mechanical ventilation) and multi-organ dysfunction (Falasca et al., 2020; Lucas et al., 2020; Mehta et al., 2020; Z. Wang et al., 2020).

The acquired immune responses of SARS-CoV-2 are related to reduced severity of COVID-19. The minimum or absent immunity is related to severe infection of COVID-19. Notable compensation may survive between the protective actions of specific CD4 T cells, specific CD8 T cells and neutralising antibodies (Moderbacher et al., 2020).

The CD4 + T lymphocytes are rapidly activated to become pathogenic T helper (Th) 1 cell and generate granulocyte-macrophage colony-stimulating factor (GM-CSF). Also, over-activation of T cells, manifested by the increase in Th17 and high cytotoxicity of CD8 + T cells in the peripheral blood of a patient with severe COVID-19, have been reported (Xu et al., 2020). Although the pathophysiology of SARS-CoV-2 is not yet fully understood, it seems there are similarities with that of SARS-CoV-1 (Crisan-Dabija et al., 2020).

Some therapeutics are under consideration for the early stages of COVID-19 infection, including Interferon alpha, vitamin B3, zinc, convalescent plasma therapy and antivirals, such as remdesivir, favipiravir, casirivimab/imdevimab antibody cocktail and bamlanivimab (Holshue et al., 2020; M. Wang et al., 2020). Nonetheless, hydroxychloroquine and lopinavir/ritonavir drugs have not been strongly associated with dissimilarity in hospital mortality (Cortegiani et al., 2021; Rosenberg et al., 2020). In severe cases, dexamethasone is being endorsed for usage and hyaluronan synthase 2, activated mesenchymal stem cells are being evaluated (Cantini, Niccoli, Matarrese, et al., 2020; Cantini, Niccoli, Nannini, et al., 2020; Mehta et al., 2020). Respiratory infections with virus and TB damage the immune system of the host, but there is very limited proof of COVID-19 and TB co-infection. The active TB infection might stimulate the progression of COVID-19 infection and aggravate the course of disease in the coinfected individuals (Yasri & Wiwanitkit, 2020) and the evidence provided by a study performed on a systematic transcriptomic evaluation of immune signatures associated with COVID-19 clinical severity and the spectrum of asymptomatic and symptomatic TB. A study of transcriptomic analysis suggested that active TB cases enhance the risk of severity of COVID-19 disease because of the excessive circulating myeloid subpopulations which are also found in the lungs of severe COVID-19 patients (Sheerin et al., 2020). The interferons (IFN) (type I and III IFN) are upregulated in both COVID-19 (Acharya et al., 2020) and TB infections (Cliff et al., 2015) and may lead to the progress and severity of the disease. Therefore, the COVID-19 disease possesses the biggest menace to ending the TB epidemic.

4.4 DIAGNOSTIC APPROACHES FOR COVID-19 AND TUBERCULOSIS

4.4.1 Diagnostic Approaches for COVID-19

After the emergence of COVID-19, immense efforts have been taken for the development of diagnostic approaches for COVID-19 infected patients (Falzone et al., 2021).

Many of the diagnostic methods have been endorsed till date, but it is still difficult to use the right test for the accurate identification according to the patient's medical history (Sanyaolu et al., 2020). Many parameters should be kept in mind while selecting of the optimal and proper diagnostic test, such as sensitivity, specificity, rapidity, cost and availability of the technologies at the health care centres (Falzone et al., 2021). This chapter provides the details of tests for COVID-19 diagnosis available on the market:

4.4.1.1 RT-PCR Test

The reverse transcriptase (RT)-PCR-based molecular methods are the gold standard techniques and are advised as the optimal diagnostic approaches due to the low cost of the kits for the final diagnosis of the COVID-19 infection (Mahendiratta et al., 2020). Many of the genes are being used for designing the primer and probes, such as RdRP gene, Nucleocapsid (N gene), Spike molecules (S gene), Envelope (E gene), etc. RT-PCR methods are time-saving and easy techniques and do not need highly trained personnel (Neilan et al., 2020). Summarily, nasopharyngeal/oropharyngeal swabs are collected from suspected patients of COVID-19 infection, and the viral RNA is extracted using commercial kits. Later, the RNA is used as a template in the RT-PCR protocol. The PCR can be performed as a one-step and two-step RT-PCR kit. After this, targets are amplified by RT-PCR using specific primers and probes. The enzyme Taq polymerase possesses the 5′-3′ exonuclease activity through which it cuts the probes attached at the template strand. The removal of the probes allows the fluorescence detection, which corresponds to the amplification of the template. The whole procedure of RT-PCR takes six to eight hours. This test is rapid, sensitive and highly specific, and the instruments are easily available in hospitals and laboratories for the diagnosis of COVID-19 (Falzone et al., 2021).

4.4.1.2 Rapid Antibody and Antigen Tests

To monitor and manage the rapidly increasing cases of COVID-19 (Augustine et al., 2020; Cerutti et al., 2020; Olalekan et al., 2020), many tests have been developed for the detection of anti-SARS-CoV-2 human antibodies in saliva, blood, nasal or oropharyngeal swab samples (Albert et al., 2021). In comparison with RT-PCR, rapid antigen-antibody tests are cost-effective and give results within ~15–30 minutes (Pilarowski et al., 2021). Rapid antigen and antibody tests are built on the platform-based technology based on lateral flow immunoassay (LFIA). In rapid antigen test the SARS-CoV-2 are targeted by loading the swab sample into the sample well of the cartridge, and it reaches on the conjugation pad immobilised with specific antibodies against the SARS-CoV-2 virus (Diao et al., 2021; J.-H. Lee et al., 2021).

The rapid antibody test targets the IgA or IgG and IgM human antibody against SARS-CoV-2 antigens. To perform the test, the swab sample is loaded into the sample well, and it reaches to the conjugated pad containing gold-tagged viral antigens and control antibodies. The anti-SARS-CoV-2 antibodies bind to the gold-tagged antigens and further attach to membrane. The test gives a colourimetric reaction in case of positive reaction in test and control zone (Falzone et al., 2021). In the case of negativity, only the control band is coloured. If no band is observed or the test band is observed but the control band is absent, the test is considered as invalid (Prazuck et al., 2020). The antigen and antibody test can be performed as point-of-care testing and gives results in a very short period without the use of specific instruments (Lanser et al., 2021; Mak et al., 2020). These tests suffer from important limitations mainly related to low sensitivity and specificity of 56.2% and 99.5%, respectively (Dinnes et al., 2020).

4.4.1.3 Serological Tests

The fluorescent and microwell plate based indirect ELISA assay is the basis of the serological tests. ELISA is performed for the detection and quantitation of proteins, antibodies, peptides and antigens and gives results in one to five hours with high sensitivity and specificity (Falzone et al., 2021).

For antibody detection: The 96-well commercial indirect ELISA assay for COVID-19 contains viral antigen attached at the bottom of the wells, which binds to the anti-SARS-CoV-2 antibodies in the serum sample. The washing steps are carried out to remove the excess antibodies and serum sample. Later on, human immunoglobulins specific conjugated antibody is added to each of the wells. More washing steps are performed followed by adding a chromogenic substrate. The substrate is metabolised and gives a colour reaction detected by optical densitometry which indicates the quantity of IgM (or IgA), IgG, antibodies present (Falzone et al., 2021).

For antigen detection: ELISA assay contains antibodies attached at the bottom of wells which binds to the viral antigens in the serum sample. In this assay SARS-CoV-2 specific conjugated antibody and unbounded antigens are added to each of the well. Remaining steps of washing, substrate addition are the same as the antibody detection method. This assay was developed for surveillance and monitoring the immunological status of patients (Alharbi et al., 2020; Korth et al., 2020).

Summarily, ELISA assay is a good clinical option for surveillance and accurate quantitation and identification of the antigens and antibodies. Also, the test can be performed for multiple samples in one go (Falzone et al., 2021).

4.4.2 Smear Microscopy using Ziehl Neelsen (ZN) Staining Method

A smear is prepared and stained by heated carbol fuchsin to solubilise the lipid material found in the cell wall of mycobacteria. Then, the smear is decolourised with a decoloriser. The lipid is absent in the cell wall of non-acid-fast bacilli, so they get decolourised easily. Then the smear is stained with methylene blue. The non-acid-fast cells absorb methylene blue easily and show the blue colour and acid-fast cells retain the red colour (Aryal, 2015). The slide is observed under the light microscope (Maurya, 2017). The acid-fast bacilli (AFB) gives red or purple colour and the non-acid-fast cells show a blue colour.

4.4.2.1 Fluorescent Microscopy

The rhodamine and auramine dye is used in this method which binds with the mycolic acids in the cell wall. AFB fluoresces under ultraviolet light and appears to be yellow or orange in colour. This staining method is more sensitive but non-specific for the detection of mycobacteria (Girma et al., 2018).

4.4.2.2 TB Culture

The culture of M. tuberculosis which is considered a gold standard is performed using the solid and liquid culture medium. Lowenstein-Jensen (LJ) medium which is a solid culture medium takes eight weeks of incubation and Middlebrook liquid medium which takes six to eight weeks for growth is used for the cultivation of both M. tuberculosis (Griffith et al., 2007). Cultivation using the liquid medium needs an antibiotic mixture named PANTA (polymyxin B, amphotericin B, nalidixic acid, trimethoprim and azlocillin) for inhibition of other contaminants and a supplement mixture named OADC (oleic acid, bovine albumin, dextrose and catalase) to increase the growth (Carricajo et al., 2001; Conville et al., 1995; Somoskövi & Magyar, 1999; Whittier et al., 1993).

4.4.2.3 Immunodiagnostics

The immunodiagnostics can be divided into (i) Interferon-γ release assays [IGRA] and tuberculin skin test (TST) for the detection of latent TB infection; (ii) Humoral immune response-based antibody detection (serological) assays which are intended to diagnose active TB disease.

4.4.2.4 Interferon Gamma Release Assay

A study in recent times concluded that IGRA can give an advantage over TST (Arend et al., 2000; Brock et al., 2001; Lalvani et al., 2001; Ravn et al., 1999). The commercially available IGRA kits are Quantiferon TB Kit and enzyme-linked immunospot spot TB assay.

4.4.2.5 Immunochromatographic Test (ICT)

The ICT was developed to upgrade the diagnosis of TB. It is based on the detection of the antibody secreted for the purified antigens of MTB. The SD Bioline MPT-64 TB Ag kit is a rapid and simple test used for the rapid identification and differentiation of MTB and NTM with a reported sensitivity and specificity of 99% and 100% (Andersen et al., 1991; Ismail et al., 2009; Li et al., 1993; Maurya et al., 2012; Nagai et al., 1991; Steingart et al., 2009).

4.4.2.6 Nucleic acid Amplification Test (NAAT)

The development of this test is very crucial and important for the diagnosis of TB with commercial availability over the period of 20 years in the USA with more accuracy as compared to the conventional techniques (Davis et al., 2014; Marks et al., 2013).

4.4.2.6.1 Xpert MTB/RIF Assay

This test was developed by Cepheid Inc., Sunnyvale, California, USA (World Health Organization, 2013), it is a simple, fast, very sensitive molecular-based (PCR) test that detects TB and resistance to rifampicin drug (Blakemore et al., 2010). The WHO endorsed t he use of this test in the year 2010, and it is a fully automated cartridge-based test that gives results in less than 60 minutes (Iram et al., 2015; World Health Organization, 2014).

4.4.2.6.2 Truenat MTB Assay

The new, chip-based molecular test was developed by Molbio Diagnostics, India and endorsed by WHO in 2019 for the detection of TB and drug resistance of rifampicin (Nikam et al., 2014). The battery-operated devices are used for extraction (Trueprep AUTO device) and amplification (TrueNat MTB chip) of the DNA. The analysis and reading part is done by the TrueLab PCR analyser, and the test gives results in approximately 60 minutes and an additional 60 minutes for rifampicin drug resistance testing. TrueNat reduces the turnaround time, so it could help in the treatment initiation (D. J. Lee et al., 2019).

4.4.2.7 Loop Mediated Isothermal Amplification

This test was developed by Eiken Chemical Co. Ltd., Tokyo, Japan, and is an isothermal nucleic acid amplification test (Notomi et al., 2000), in which amplification takes place under isothermal conditions. It has been successfully executed in clinical settings as a screening tool (Parida et al., 2008). The WHO endorsed this test in August 2016 for the diagnosis of pulmonary TB in adults. Many studies have revealed that the LAMP assay is specific, fast, very simple and inexpensive compared to PCR-based tests (Bentaleb et al., 2016).

4.5 IMPACT OF COVID-19 ON TB

COVID-19 has direct and indirect consequences on the economy which will affect the global TB programmes. The likely impact of COVID-19 on TB control is diagnosis and treatment therapy are hampered, poor after-effects of drugs, escalation in TB transmission and the threat of multi-drug-resistant TB. COVID-19 is spread by aerosol transmission so the prevention measures taken by authorities are advising people to stay at home to lessen the community transmission till the situation comes under control (Anderson et al., 2020). This preventive measure promotes the transmission of TB since the very reason for TB transmission is continuous contact in a household (Acuña-Villaorduña et al., 2018). A study by Cilloni et al. (2020) revealed that the COVID-19 lockdown for a period of 90 days would give rise to an extra 1.6–1.7 million TB cases and around 438,000 TB deaths in India over the period of the next five years. The after-effect of the household transmission of TB will be declared only in the future when the cases will increase because TB has a prolonged incubation period (McCreesh & White, 2018; Ragonnet et al., 2019). This suggests that the observation of public health is advisable.

The COVID-19 cases have an impact on the routine diagnostic services of TB as well. In a pandemic situation, there was an urgent need for human resources and finance, so the resources shifted to manage COVID-19. The attention of health care services, the public, politicians and government authorities shifted to pandemic management and accountability for TB management was reduced. Overburden of the health care staff leads to error and poor quality control of the tests. The non-availability of the health care staff was due to medical conditions. Fear of COVID-19 in public stops them from visiting the hospital and DOTS centre. The above-mentioned factors will come up with a delayed diagnosis and start of the treatment which will enhance the transmission risk of TB, poor outcomes and expansion of drug-resistant TB (Comas et al., 2013; Kyu et al., 2018; "The global burden of tuberculosis," 2018; Vynnycky & Fine, 1999).

Impact of COVID-19 on the prevention and control of TB: The programmes for TB Prevention and control have already suffered because of the pandemic. The exchanges of information through seminars, conferences and workshops have stopped. Vaccination programmes for TB and the global strategy of ending TB by 2035 have been affected worldwide due to COVID-19. The COVID-19 infection causes a weak immune system and respiratory failure which might be related to the risk of developing TB (Qin et al., 2020; Shi et al., 2020; Singh et al., 2012). Studies have revealed that virus infections illustrate the evolution of active TB disease (Barnes et al., 1991; Noymer, 2008; Pawlowski et al., 2012).

Besides, the poor economy will show an excessive effect on the lower class via disturbance in the education of children, health care services and loss of jobs and money (Sumner et al., 2020). As per the estimation of the World Bank, the global poverty rate could increase by 0.3–0.7 % to haract 09 % in the year 2020, and 40–60 million individuals will face poverty in the year 2020 due to the pandemic condition. These conditions will impact TB notifications and control programmes worldwide (Oxlade & Murray, 2012; Spence et al., 1993).

4.6 CONCLUSION

COVID-19 disease can cause asymptomatic to severe infection which can be fatal. The use of immunosuppressants during the COVID-19 infection may result in the reactivation of TB. A person can be infected with COVID-19 infection at any time during the course of TB infection. More studies need to be done to understand the TB and COVID-19 co-infection. The common symptoms and signs of both diseases are the same. The available literature and research are limited to knowing the effect of COVID-19 on TB patients with ongoing treatment. The main factors associated with the mortality in COVID-19 cases are age, comorbidities such as diabetes and HIV. In concluding remarks, we need more studies to understand the questions raised in this chapter. Till then people should follow the strict guidelines and get vaccinated for COVID-19, and patients with active TB disease should try to not get COVID-19 infection.

REFERENCES

Acharya, D., Liu, G., & Gack, M. U. (2020). Dysregulation of type I interferon responses in COVID-19. *Nature Reviews Immunology*, *20*(7), 397–398. https://doi.org/10.1038/s41577-020-0346-x.

Acuña-Villaorduña, C., Jones-Lopez, E., Fregona, G., Marques-Rodriguez, P., Geadas, C., Hadad, D. J., White, L. F., Molina, L. P. D., Vinhas, S., Gaeddert, M., Ribeiro-Rodriguez, R., Salgame, P., Palaci, M., Alland, D., Ellner, J. J., & Dietze, R. (2018). Intensity of exposure to pulmonary tuberculosis determines risk of tuberculosis infection and disease. *The European Respiratory Journal*, *51*(1), 1701578. https://doi.org/10.1183/13993003.01578-2017.

Albert, E., Torres, I., Bueno, F., Huntley, D., Molla, E., Fernández-Fuentes, M. Á., Martínez, M., Poujois, S., Forqué, L., Valdivia, A., Solano de la Asunción, C., Ferrer, J., Colomina, J., & Navarro, D. (2021). Field evaluation of a rapid antigen test (Panbio™ COVID-19 Ag Rapid Test Device) for COVID-19 diagnosis in primary healthcare centres. *Clinical Microbiology and Infection*, *27*(3), 472.e7–472.e10. https://doi.org/10.1016/j.cmi.2020.11.004.

Alene, K. A., Wangdi, K., & Clements, A. C. A. (2020). Impact of the COVID-19 pandemic on tuberculosis control: An overview. *Tropical Medicine and Infectious Disease*, *5*(3), 123. https://doi.org/10.3390/tropicalmed5030123.

Alharbi, S. A., Almutairi, A. Z., Jan, A. A., & Alkhalify, A. M. (2020). Enzyme-linked immunosorbent assay for the detection of severe acute respiratory syndrome coronavirus 2 (SARS-CoV-2) IgM/IgA and IgG antibodies among healthcare workers. *Cureus*, *12*(9), e10285. https://doi.org/10.7759/cureus.10285.

Andersen, P., Askgaard, D., Ljungqvist, L., Bennedsen, J., & Heron, I. (1991). Proteins released from Mycobacterium tuberculosis during growth. *Infection and Immunity*, *59*(6), 1905–1910.

Anderson, R. M., Heesterbeek, H., Klinkenberg, D., & Hollingsworth, T. D. (2020). How will country-based mitigation measures influence the course of the COVID-19 epidemic? *Lancet (London, England)*, *395*(10228), 931–934. https://doi.org/10.1016/S0140-6736(20)30567-5.

Arend, S. M., Andersen, P., van Meijgaarden, K. E., Skjot, R. L., Subronto, Y. W., van Dissel, J. T., & Ottenhoff, T. H. (2000). Detection of active tuberculosis infection by T cell responses to early-secreted antigenic target 6-kDa protein and culture filtrate protein 10. *The Journal of Infectious Diseases*, *181*(5), 1850–1854. https://doi.org/10.1086/315448.

Aryal, S. (May 8, 2015). Acid-fast stain-principle, procedure, interpretation and examples. *Microbiology Info.Com*. https://microbiologyinfo.com/acid-fast-stain-principle-procedure-interpretation-and-examples/.

Augustine, R., Das, S., Hasan, A. S. A., Abdul Salam, S., Augustine, P., Dalvi, Y. B., Varghese, R., Primavera, R., Yassine, H. M., Thakor, A. S., & Kevadiya, B. D. (2020). Rapid antibody-based COVID-19 mass surveillance: Relevance, challenges, and prospects in a pandemic and post-pandemic world. *Journal of Clinical Medicine*, *9*(10), 3372. https://doi.org/10.3390/jcm9103372.

Aznar, M. L., Espinosa-Pereiro, J., Saborit, N., Jové, N., Sánchez Martinez, F., Pérez-Recio, S., Vitoria, A., Sanjoaquin, I., Gallardo, E., Llenas-García, J., Pomar, V., García, I. O., Cacho, J., Goncalves De Freitas, L., San Martin, J. V., García Rodriguez, J. F., Jiménez-Fuentes, M. Á., De Souza-Galvao, M. L., Tórtola, T., . . . Sánchez-Montalvá, A. (2021). Impact of the COVID-19 pandemic on tuberculosis management in Spain. *International Journal of Infectious Diseases*, *108*, 300–305. https://doi.org/10.1016/j.ijid.2021.04.075.

Barnes, P. F., Bloch, A. B., Davidson, P. T., & Snider, D. E. (1991). Tuberculosis in patients with human immunodeficiency virus infection. *The New England Journal of Medicine*, *324*(23), 1644–1650. https://doi.org/10.1056/NEJM199106063242307.

Bentaleb, E. M., Abid, M., El Messaoudi, M. D., Lakssir, B., Ressami, E. M., Amzazi, S., Sefrioui, H., & Ait Benhassou, H. (2016). Development and evaluation of an in-house single step loop-mediated isothermal amplification (SS-LAMP) assay for the detection of Mycobacterium tuberculosis complex in sputum samples from Moroccan patients. *BMC Infectious Diseases*, *16*(1), 517. https://doi.org/10.1186/s12879-016-1864-9.

Blakemore, R., Story, E., Helb, D., Kop, J., Banada, P., Owens, M. R., Chakravorty, S., Jones, M., & Alland, D. (2010). Evaluation of the analytical performance of the Xpert MTB/RIF assay. *Journal of Clinical Microbiology*, *48*(7), 2495–2501. https://doi.org/10.1128/JCM.00128-10.

Brock, I., Munk, M. E., Kok-Jensen, A., & Andersen, P. (2001). Performance of whole blood IFN-gamma test for tuberculosis diagnosis based on PPD or the specific antigens ESAT-6 and CFP-10. *The International Journal of Tuberculosis and Lung Disease: The Official Journal of the International Union Against Tuberculosis and Lung Disease*, *5*(5), 462–467.

Cantini, F., Niccoli, L., Matarrese, D., Nicastri, E., Stobbione, P., & Goletti, D. (2020). Baricitinib therapy in COVID-19: A pilot study on safety and clinical impact. *Journal of Infection*, *81*(2), 318–356. https://doi.org/10.1016/j.jinf.2020.04.017.

Cantini, F., Niccoli, L., Nannini, C., Matarrese, D., Natale, M. E. D., Lotti, P., Aquilini, D., Landini, G., Cimolato, B., Pietro, M. A. D., Trezzi, M., Stobbione, P., Frausini, G., Navarra, A., Nicastri, E., Sotgiu, G., & Goletti, D. (2020). Beneficial impact of Baricitinib in COVID-19 moderate pneumonia: Multicentre study. *Journal of Infection*, *81*(4), 647–679. https://doi.org/10.1016/j.jinf.2020.06.052.

Carricajo, A., Fonsale, N., Vautrin, A. C., & Aubert, G. (2001). Evaluation of BacT/Alert 3D liquid culture system for recovery of mycobacteria from clinical specimens using sodium dodecyl (lauryl) sulfate-NaOH decontamination. *Journal of Clinical Microbiology*, *39*(10), 3799–3800. https://doi.org/10.1128/JCM.39.10.3799-3800.2001.

Cerutti, F., Burdino, E., Milia, M. G., Allice, T., Gregori, G., Bruzzone, B., & Ghisetti, V. (2020). Urgent need of rapid tests for SARS CoV-2 antigen detection: Evaluation of the SD-Biosensor antigen test for SARS-CoV-2. *Journal of Clinical Virology*, *132*, 104654. https://doi.org/10.1016/j.jcv.2020.104654.

Cilloni, L., Fu, H., Vesga, J. F., Dowdy, D., Pretorius, C., Ahmedov, S., Nair, S. A., Mosneaga, A., Masini, E., Sahu, S., & Arinaminpathy, N. (2020). The potential impact of the COVID-19 pandemic on the tuberculosis epidemic a modelling analysis. *EClinicalMedicine*, *28*, 100603. https://doi.org/10.1016/j.eclinm.2020.100603

Cliff, J. M., Kaufmann, S. H. E., McShane, H., van Helden, P., & O'Garra, A. (2015). The human immune response to tuberculosis and its treatment: A view from the blood. *Immunological Reviews*, *264*(1), 88–102. https://doi.org/10.1111/imr.12269.

Comas, I., Coscolla, M., Luo, T., Borrell, S., Holt, K. E., Kato-Maeda, M., Parkhill, J., Malla, B., Berg, S., Thwaites, G., Yeboah-Manu, D., Bothamley, G., Mei, J., Wei, L., Bentley, S., Harris, S. R., Niemann, S., Diel, R., Aseffa, A., . . . Gagneux, S. (2013). Out-of-Africa migration and neolithic co-expansion of Mycobacterium tuberculosis with modern humans. *Nature Genetics*, *45*(10), 1176–1182. https://doi.org/10.1038/ng.2744.

Conville, P. S., Andrews, J. W., & Witebsky, F. G. (1995). Effect of PANTA on growth of Mycobacterium kansasii in BACTEC 12B medium. *Journal of Clinical Microbiology*, *33*(8), 2012–2015. https://doi.org/10.1128/jcm.33.8.2012-2015.1995.

Cortegiani, A., Ippolito, M., Greco, M., Granone, V., Protti, A., Gregoretti, C., Giarratano, A., Einav, S., & Cecconi, M. (2021). Rationale and evidence on the use of tocilizumab in COVID-19: A systematic review. *Pulmonology*, *27*(1), 52–66. https://doi.org/10.1016/j.pulmoe.2020.07.003.

Crisan-Dabija, R., Grigorescu, C., Pavel, C.-A., Artene, B., Popa, I. V., Cernomaz, A., & Burlacu, A. (2020). Tuberculosis and COVID-19: Lessons from the past viral outbreaks and possible future outcomes. *Canadian Respiratory Journal*, *2020*, 1401053. https://doi.org/10.1155/2020/1401053.

Davis, J. L., Kawamura, L. M., Chaisson, L. H., Grinsdale, J., Benhammou, J., Ho, C., Babst, A., Banouvong, H., Metcalfe, J. Z., Pandori, M., Hopewell, P. C., & Cattamanchi, A. (2014). Impact of GeneXpert MTB/RIF on patients and tuberculosis programs in a low-Burden setting. A hypothetical trial. *American Journal of Respiratory and Critical Care Medicine*, *189*(12), 1551–1559. https://doi.org/10.1164/rccm.201311-1974OC.

Diao, B., Wen, K., Zhang, J., Chen, J., Han, C., Chen, Y., Wang, S., Deng, G., Zhou, H., & Wu, Y. (2021). Accuracy of a nucleocapsid protein antigen rapid test in the diagnosis of SARS-CoV-2 infection. *Clinical Microbiology and Infection*, *27*(2), 289.e1–289. https://doi.org/10.1016/j.cmi.2020.09.057.

Dinnes, J., Deeks, J. J., Adriano, A., Berhane, S., Davenport, C., Dittrich, S., Emperador, D., Takwoingi, Y., Cunningham, J., Beese, S., Dretzke, J., Ferrante di Ruffano, L., Harris, I. M., Price, M. J., Taylor-Phillips, S., Hooft, L., Leeflang, M. M., Spijker, R., Van den Bruel, A., & Cochrane COVID-19 Diagnostic Test Accuracy Group. (2020). Rapid, point-of-care antigen and molecular-based tests for diagnosis of SARS-CoV-2 infection. *The Cochrane Database of Systematic Reviews*, *8*, CD013705. https://doi.org/10.1002/14651858.CD013705.

Falasca, L., Nardacci, R., Colombo, D., Lalle, E., Di Caro, A., Nicastri, E., Antinori, A., Petrosillo, N., Marchioni, L., Biava, G., D'Offizi, G., Palmieri, F., Goletti, D., Zumla, A., Ippolito, G., Piacentini, M., Del Nonno, F., & COVID-19 INMI Study Group. (2020). Postmortem findings in Italian patients with COVID-19: A descriptive full autopsy study of cases with and without comorbidities. *The Journal of Infectious Diseases*, *222*(11), 1807–1815. https://doi.org/10.1093/infdis/jiaa578.

Falzone, L., Gattuso, G., Tsatsakis, A., Spandidos, D. A., & Libra, M. (2021). Current and innovative methods for the diagnosis of COVID-19 infection (Review). *International Journal of Molecular Medicine*, *47*(6), 100. https://doi.org/10.3892/ijmm.2021.4933.

Gao, Y., Liu, M., Chen, Y., Shi, S., Geng, J., & Tian, J. (2021). Association between tuberculosis and COVID-19 severity and mortality: A rapid systematic review and meta-analysis. *Journal of Medical Virology*, *93*(1), 194–196. https://doi.org/10.1002/jmv.26311.

Girma, S., Avanzi, C., Bobosha, K., Desta, K., Idriss, M. H., Busso, P., Tsegaye, Y., Nigusse, S., Hailu, T., Cole, S. T., & Aseffa, A. (2018). Evaluation of auramine O staining and conventional PCR for leprosy diagnosis: A comparative cross-sectional study from Ethiopia. *PloS Neglected Tropical Diseases*, *12*(9), e0006706. https://doi.org/10.1371/journal.pntd.0006706.

Global Tuberculosis report 2021. (n.d.). Retrieved June 25, 2023, from https://www.who.int/publications-detail-redirect/9789240037021

Gorbalenya, A. E., Baker, S. C., Baric, R. S., de Groot, R. J., Drosten, C., Gulyaeva, A. A., . . . & Ziebuhr, J. (2020). The species severe acute respiratory syndrome-related coronavirus: Classifying 2019-nCoV and naming it SARS-CoV-2. *Nature Microbiology*, *5*(4), 536–544.

Griffith, D. E., Aksamit, T., Brown-Elliott, B. A., Catanzaro, A., Daley, C., Gordin, F., Holland, S. M., Horsburgh, R., Huitt, G., Iademarco, M. F., Iseman, M., Olivier, K., Ruoss, S., von Reyn, C. F., Wallace, R. J., Winthrop, K., & ATS Mycobacterial Diseases Subcommittee, American Thoracic Society, & Infectious Disease Society of America. (2007). An official ATS/IDSA statement: Diagnosis, treatment, and prevention of nontuberculous mycobacterial diseases. *American Journal of Respiratory and Critical Care Medicine*, *175*(4), 367–416. https://doi.org/10.1164/rccm.200604-571ST.

Gupta, N., Ish, P., Gupta, A., Malhotra, N., Caminero, J. A., Singla, R., Kumar, R., Yadav, S. R., Dev, N., Agrawal, S., Kohli, S., Sen, M. K., Chakrabarti, S., & Gupta, N. K. (2020). A profile of a retrospective cohort of 22 patients with COVID-19 and active/treated tuberculosis. *European Respiratory Journal*, *56*(5). https://doi.org/10.1183/13993003.03408-2020.

Hanna, T. P., Evans, G. A., & Booth, C. M. (2020). Cancer, COVID-19 and the precautionary principle: Prioritizing treatment during a global pandemic. *Nature Reviews. Clinical Oncology*, *17*(5), 268–270. https://doi.org/10.1038/s41571-020-0362-6.

Hoffmann, M., Kleine-Weber, H., Schroeder, S., Krüger, N., Herrler, T., Erichsen, S., Schiergens, T. S., Herrler, G., Wu, N.-H., Nitsche, A., Müller, M. A., Drosten, C., & Pöhlmann, S. (2020). SARS-CoV-2 cell entry depends on ACE2 and TMPRSS2 and is blocked by a clinically proven protease inhibitor. *Cell*, *181*(2), 271–280. https://doi.org/10.1016/j.cell.2020.02.052.

Holshue, M. L., DeBolt, C., Lindquist, S., Lofy, K. H., Wiesman, J., Bruce, H., Spitters, C., Ericson, K., Wilkerson, S., Tural, A., Diaz, G., Cohn, A., Fox, L., Patel, A., Gerber, S. I., Kim, L., Tong, S., Lu, X., Lindstrom, S., . . . Pillai, S. K. (2020). First case of 2019 novel coronavirus in the United States. *New England Journal of Medicine*, *382*(10), 929–936. https://doi.org/10.1056/NEJMoa2001191.

Ippolito, M., Vitale, F., Accurso, G., Iozzo, P., Gregoretti, C., Giarratano, A., & Cortegiani, A. (2020). Medical masks and respirators for the protection of healthcare workers from SARS-CoV-2 and other viruses. *Pulmonology*, *26*(4), 204–212. https://doi.org/10.1016/j.pulmoe.2020.04.009.

Iram, S., Zeenat, A., Hussain, S., Wasim Yusuf, N., & Aslam, M. (2015). Rapid diagnosis of tuberculosis using Xpert MTB/RIF assay—report from a developing country. *Pakistan Journal of Medical Sciences*, *31*(1), 105–110. https://doi.org/10.12669/pjms.311.6970.

Ismail, N. A., Baba, K., Pombo, D., & Hoosen, A. A. (2009). Use of an immunochromatographic kit for the rapid detection of Mycobacterium tuberculosis from broth cultures. *The International Journal of Tuberculosis and Lung Disease: The Official Journal of the International Union Against Tuberculosis and Lung Disease*, *13*(8), 1045–1047.

Korth, J., Wilde, B., Dolff, S., Anastasiou, O. E., Krawczyk, A., Jahn, M., Cordes, S., Ross, B., Esser, S., Lindemann, M., Kribben, A., Dittmer, U., Witzke, O., & Herrmann, A. (2020). SARS-CoV-2-specific antibody detection in healthcare workers in Germany with direct contact to COVID-19 patients. *Journal of Clinical Virology: The Official Publication of the Pan American Society for Clinical Virology*, *128*, 104437. https://doi.org/10.1016/j.jcv.2020.104437.

Kyu, H. H., Maddison, E. R., Henry, N. J., Mumford, J. E., Barber, R., Shields, C., Brown, J. C., Nguyen, G., Carter, A., Wolock, T. M., Wang, H., Liu, P. Y., Reitsma, M., Ross, J. M., Abajobir, A. A., Abate, K. H., Abbas, K., Abera, M., Abera, S. F., . . . Murray, C. J. (2018). The global burden of tuberculosis: Results from the global burden of disease study 2015. *The Lancet Infectious Diseases*, *18*(3), 261–284. https://doi.org/10.1016/S1473-3099(17)30703-X.

Lalvani, A., Pathan, A. A., McShane, H., Wilkinson, R. J., Latif, M., Conlon, C. P., Pasvol, G., & Hill, A. V. (2001). Rapid detection of Mycobacterium tuberculosis infection by enumeration of antigen-specific T cells. *American Journal of Respiratory and Critical Care Medicine*, *163*(4), 824–828. https://doi.org/10.1164/ajrccm.163.4.2009100.

Lanser, L., Bellmann-Weiler, R., Öttl, K.-W., Huber, L., Griesmacher, A., Theurl, I., & Weiss, G. (2021). Evaluating the clinical utility and sensitivity of SARS-CoV-2 antigen testing in relation to RT-PCR Ct values. *Infection*, *49*(3), 555–557. https://doi.org/10.1007/s15010-020-01542-0.

Lee, D. J., Kumarasamy, N., Resch, S. C., Sivaramakrishnan, G. N., Mayer, K. H., Tripathy, S., Paltiel, A. D., Freedberg, K. A., & Reddy, K. P. (2019). Rapid, point-of-care diagnosis of tuberculosis with novel Truenat assay: Cost-effectiveness analysis for India's public sector. *PloS One*, *14*(7), e0218890. https://doi.org/10.1371/journal.pone.0218890.

Lee, J.-H., Choi, M., Jung, Y., Lee, S. K., Lee, C.-S., Kim, J., Kim, J., Kim, N. H., Kim, B.-T., & Kim, H. G. (2021). A novel rapid detection for SARS-CoV-2 spike 1 antigens using human angiotensin converting enzyme 2 (ACE2). *Biosensors & Bioelectronics*, *171*, 112715. https://doi.org/10.1016/j.bios.2020.112715.

Li, H., Ulstrup, J. C., Jonassen, T. O., Melby, K., Nagai, S., & Harboe, M. (1993). Evidence for absence of the MPB64 gene in some substrains of Mycobacterium bovis BCG. *Infection and Immunity*, *61*(5), 1730–1734. https://doi.org/10.1128/iai.61.5.1730-1734.1993.

Lu, R., Zhao, X., Li, J., Niu, P., Yang, B., Wu, H., Wang, W., Song, H., Huang, B., Zhu, N., Bi, Y., Ma, X., Zhan, F., Wang, L., Hu, T., Zhou, H., Hu, Z., Zhou, W., Zhao, L., . . . Tan, W. (2020a). Genomic characterisation and epidemiology of 2019 novel coronavirus: Implications for virus origins and receptor binding. *Lancet (London, England)*, *395*(10224), 565–574. https://doi.org/10.1016/S0140-6736(20)30251-8.

Lu, R., Zhao, X., Li, J., Niu, P., Yang, B., Wu, H., Wang, W., Song, H., Huang, B., Zhu, N., Bi, Y., Ma, X., Zhan, F., Wang, L., Hu, T., Zhou, H., Hu, Z., Zhou, W., Zhao, L., . . . Tan, W. (2020b). Genomic characterisation and epidemiology of 2019 novel coronavirus: Implications for virus origins and receptor binding. *The Lancet*, *395*(10224), 565–574. https://doi.org/10.1016/S0140-6736(20)30251-8.

Lucas, C., Wong, P., Klein, J., Castro, T. B. R., Silva, J., Sundaram, M., Ellingson, M. K., Mao, T., Oh, J. E., Israelow, B., Takahashi, T., Tokuyama, M., Lu, P., Venkataraman, A., Park, A., Mohanty, S., Wang, H., Wyllie, A. L., Vogels, C. B. F., . . . Iwasaki, A. (2020). Longitudinal analyses reveal immunological misfiring in severe COVID-19. *Nature*, *584*(7821), 463–469. https://doi.org/10.1038/s41586-020-2588-y.

Maddaloni, E., & Buzzetti, R. (2020). Covid-19 and diabetes mellitus: Unveiling the interaction of two pandemics. *Diabetes/Metabolism Research and Reviews*, e33213321. https://doi.org/10.1002/dmrr.3321.

Mahendiratta, S., Batra, G., Sarma, P., Kumar, H., Bansal, S., Kumar, S., Prakash, A., Sehgal, R., & Medhi, B. (2020). Molecular diagnosis of COVID-19 in different biologic matrix, their diagnostic validity and clinical relevance: A systematic review. *Life Sciences*, *258*, 118207. https://doi.org/10.1016/j.lfs.2020.118207.

Mak, G. C., Cheng, P. K., Lau, S. S., Wong, K. K., Lau, C. S., Lam, E. T., Chan, R. C., & Tsang, D. N. (2020). Evaluation of rapid antigen test for detection of SARS-CoV-2 virus. *Journal of Clinical Virology: The Official Publication of the Pan American Society for Clinical Virology*, *129*, 104500. https://doi.org/10.1016/j.jcv.2020.104500.

Marks, S. M., Cronin, W., Venkatappa, T., Maltas, G., Chon, S., Sharnprapai, S., Gaeddert, M., Tapia, J., Dorman, S. E., Etkind, S., Crosby, C., Blumberg, H. M., & Bernardo, J. (2013). The health-system benefits and cost-effectiveness of using Mycobacterium tuberculosis direct nucleic acid amplification testing to diagnose tuberculosis disease in the United States. *Clinical Infectious Diseases: An Official Publication of the Infectious Diseases Society of America*, *57*(4), 532–542. https://doi.org/10.1093/cid/cit336.

Maurya, A. K., Nag, V. L., Kant, S., Kushwaha, R. A. S., Kumar, M., Mishra, V., Rahman, W., & Dhole, T. N. (2012). Evaluation of an immunochromatographic test for discrimination between Mycobacterium tuberculosis complex & non tuberculous mycobacteria in clinical isolates from extra-pulmonary tuberculosis. *The Indian Journal of Medical Research*, *135*(6), 901–906.

Maurya, A. K., Nag, V. L., Kant, S., Sharma, A., Gadepalli, R. S., & Kushwaha, R. A. S. (2017). Recent methods for diagnosis of nontuberculous mycobacteria infections: Relevance in clinical practice. *Biomedical and Biotechnology Research Journal (BBRJ)*, *1*(1), 14.

McCreesh, N., & White, R. G. (2018). An explanation for the low proportion of tuberculosis that results from transmission between household and known social contacts. *Scientific Reports*, *8*, 5382. https://doi.org/10.1038/s41598-018-23797-2.

McIntosh, K., Hirsch, M. S., & Bloom, A. (2021). COVID-19: Epidemiology, virology, and prevention. *UpToDate*. Available online: https://www.uptodate.com/contents/covid-19-epidemiology-virology-and-prevention (accessed on 18 March 2022).

Mehta, P., McAuley, D. F., Brown, M., Sanchez, E., Tattersall, R. S., & Manson, J. J. (2020). COVID-19: Consider cytokine storm syndromes and immunosuppression. *The Lancet*, *395*(10229), 1033–1034. https://doi.org/10.1016/S0140-6736(20)30628-0.

Millet, J. K., & Whittaker, G. R. (2014). Host cell entry of Middle East respiratory syndrome coronavirus after two-step, Furin-mediated activation of the spike protein. *Proceedings of the National Academy of Sciences*, *111*(42), 15214–15219. https://doi.org/10.1073/pnas.1407087111.

Moderbacher, C. R., Ramirez, S. I., Dan, J. M., Grifoni, A., Hastie, K. M., Weiskopf, D., Belanger, S., Abbott, R. K., Kim, C., Choi, J., Kato, Y., Crotty, E. G., Kim, C., Rawlings, S. A., Mateus, J., Tse, L. P. V., Frazier, A., Baric, R., Peters, B., . . . Crotty, S. (2020). Antigen-specific adaptive immunity to SARS-CoV-2 in Acute COVID-19 and associations with age and disease severity. *Cell*, *183*(4), 996–1012. https://doi.org/10.1016/j.cell.2020.09.038.

Nagai, S., Wiker, H. G., Harboe, M., & Kinomoto, M. (1991). Isolation and partial characterization of major protein antigens in the culture fluid of Mycobacterium tuberculosis. *Infection and Immunity*, *59*(1), 372–382.

Neilan, A. M., Losina, E., Bangs, A. C., Flanagan, C., Panella, C., Eskibozkurt, G. E., Mohareb, A., Hyle, E. P., Scott, J. A., Weinstein, M. C., Siedner, M. J., Reddy, K. P., Harling, G., Freedberg, K. A., Shebl, F. M., Kazemian, P., & Ciaranello, A. L. (2020). Clinical impact, costs, and cost-effectiveness of expanded severe acute respiratory syndrome coronavirus 2 testing in Massachusetts. *Clinical Infectious Diseases: An Official Publication of the Infectious Diseases Society of America*, *73*(9), e2908–e2917. https://doi.org/10.1093/cid/ciaa1418.

Nikam, C., Jagannath, M., Narayanan, M. M., Ramanabhiraman, V., Kazi, M., Shetty, A., & Rodrigues, C. (2013). Rapid diagnosis of Mycobacterium tuberculosis with Truenat MTB: A near-care approach. *PLoS One*, *8*(1), e51121.

Nikam, C., Kazi, M., Nair, C., Jaggannath, M., Manoj, M., Vinaya, R., Shetty, A., & Rodrigues, C. (2014). Evaluation of the Indian TrueNAT micro RT-PCR device with GeneXpert for case detection of pulmonary tuberculosis. *International Journal of Mycobacteriology*, *3*(3), 205–210. https://doi.org/10.1016/j.ijmyco.2014.04.003.

Notomi, T., Okayama, H., Masubuchi, H., Yonekawa, T., Watanabe, K., Amino, N., & Hase, T. (2000). Loop-mediated isothermal amplification of DNA. *Nucleic Acids Research*, *28*(12), E63. https://doi.org/10.1093/nar/28.12.e63.

Noymer, A. (2008). The 1918–19 influenza pandemic affected tuberculosis in the United States: Reconsidering Bradshaw, Smith, and Blanchard. *Biodemography and Social Biology*, *54*(2), 125–133; discussion 134–140. https://doi.org/10.1080/19485565.2008.9989137.

Olalekan, A., Iwalokun, B., Akinloye, O. M., Popoola, O., Samuel, T. A., & Akinloye, O. (2020). COVID-19 rapid diagnostic test could contain transmission in low- and middle-income countries. *African Journal of Laboratory Medicine*, *9*(1), 1255. https://doi.org/10.4102/ajlm.v9i1.1255.

Oxlade, O., & Murray, M. (2012). Tuberculosis and poverty: Why are the poor at greater risk in India? *PloS One*, *7*(11), e47533. https://doi.org/10.1371/journal.pone.0047533.

Parida, M., Sannarangaiah, S., Dash, P. K., Rao, P. V. L., & Morita, K. (2008). Loop mediated isothermal amplification (LAMP): A new generation of innovative gene amplification technique: Perspectives in clinical diagnosis of infectious diseases. *Reviews in Medical Virology*, *18*(6), 407–421. https://doi.org/10.1002/rmv.593.

Parsons, L. M., Somoskövi, Á., Gutierrez, C., Lee, E., Paramasivan, C. N., Abimiku, A., Spector, S., Roscigno, G., & Nkengasong, J. (2011). Laboratory diagnosis of tuberculosis in resource-poor countries: Challenges and opportunities. *Clinical Microbiology Reviews*, *24*(2), 314–350. https://doi.org/10.1128/CMR.00059-10.

Pawlowski, A., Jansson, M., Sköld, M., Rottenberg, M. E., & Källenius, G. (2012). Tuberculosis and HIV co-infection. *PloS Pathogens*, *8*(2), e1002464. https://doi.org/10.1371/journal.ppat.1002464.

Pilarowski, G., Lebel, P., Sunshine, S., Liu, J., Crawford, E., Marquez, C., Rubio, L., Chamie, G., Martinez, J., Peng, J., Black, D., Wu, W., Pak, J., Laurie, M. T., Jones, D., Miller, S., Jacobo, J., Rojas, S., Rojas, S., . . . DeRisi, J. (2021). Performance characteristics of a rapid severe acute respiratory syndrome coronavirus 2 antigen detection assay at a public plaza testing site in San Francisco. *The Journal of Infectious Diseases*, *223*(7), 1139–1144. https://doi.org/10.1093/infdis/jiaa802.

Prasad, R., Singh, A., & Gupta, N. (2020). Tuberculosis and COVID-19 in India: Challenges and opportunities. *Lung India: Official Organ of Indian Chest Society*, *37*(4), 292–294. https://doi.org/10.4103/lungindia.lungindia_260_20.

Prazuck, T., Colin, M., Giachè, S., Gubavu, C., Seve, A., Rzepecki, V., Chevereau-Choquet, M., Kiani, C., Rodot, V., Lionnet, E., Courtellemont, L., Guinard, J., Pialoux, G., & Hocqueloux, L. (2020). Evaluation of performance of two SARS-CoV-2 rapid IgM-IgG combined antibody tests on capillary whole blood samples from the fingertip. *PloS One*, *15*(9), e0237694. https://doi.org/10.1371/journal.pone.0237694.

Qin, C., Zhou, L., Hu, Z., Zhang, S., Yang, S., Tao, Y., Xie, C., Ma, K., Shang, K., Wang, W., & Tian, D.-S. (2020). Dysregulation of immune response in patients with COVID-19 in Wuhan, China. *Clinical Infectious Diseases: An Official Publication of the Infectious Diseases Society of America*, ciaa248. https://doi.org/10.1093/cid/ciaa248.

Ragonnet, R., Trauer, J. M., Geard, N., Scott, N., & McBryde, E. S. (2019). Profiling Mycobacterium tuberculosis transmission and the resulting disease burden in the five highest tuberculosis burden countries. *BMC Medicine*, *17*, 208. https://doi.org/10.1186/s12916-019-1452-0.

Ravn, P., Demissie, A., Eguale, T., Wondwosson, H., Lein, D., Amoudy, H. A., Mustafa, A. S., Jensen, A. K., Holm, A., Rosenkrands, I., Oftung, F., Olobo, J., von Reyn, F., & Andersen, P. (1999). Human T cell responses to the ESAT-6 antigen from Mycobacterium tuberculosis. *The Journal of Infectious Diseases*, *179*(3), 637–645. https://doi.org/10.1086/314640.

Rosenberg, E. S., Dufort, E. M., Udo, T., Wilberschied, L. A., Kumar, J., Tesoriero, J., Weinberg, P., Kirkwood, J., Muse, A., DeHovitz, J., Blog, D. S., Hutton, B., Holtgrave, D. R., & Zucker, H. A. (2020). Association of treatment with hydroxychloroquine or azithromycin with in-hospital mortality in patients with COVID-19 in New York State. *JAMA*, *323*(24), 2493–2502. https://doi.org/10.1001/jama.2020.8630.

Sanyaolu, A., Okorie, C., Marinkovic, A., Ayodele, O., Abbasi, A. F., Prakash, S., Ahmed, M., Kayode, D., Jaferi, U., & Haider, N. (2020). Navigating the diagnostics of COVID-19. *Sn Comprehensive Clinical Medicine*, *2*(9), 1393–1400. https://doi.org/10.1007/s42399-020-00408-8.

Sheerin, D., Abhimanyu, W. X., Johnson, W. E., & Coussens, A. (2020). *Systematic evaluation of transcriptomic disease risk and diagnostic biomarker overlap between COVID-19 and tuberculosis: A patient-level meta-analysis* (p. 2020.11.25.20236646). medRxiv. https://doi.org/10.1101/2020.11.25.20236646.

Shi, Y., Wang, Y., Shao, C., Huang, J., Gan, J., Huang, X., Bucci, E., Piacentini, M., Ippolito, G., & Melino, G. (2020). COVID-19 infection: The perspectives on immune responses. *Cell Death and Differentiation*, *27*(5), 1451–1454. https://doi.org/10.1038/s41418-020-0530-3.

Singh, V., Sharma, B. B., & Patel, V. (2012). Pulmonary sequelae in a patient recovered from swine flu. *Lung India: Official Organ of Indian Chest Society*, *29*(3), 277–279. https://doi.org/10.4103/0970-2113.99118.

Somoskövi, A., & Magyar, P. (1999). Comparison of the mycobacteria growth indicator tube with MB redox, Löwenstein-Jensen, and Middlebrook 7H11 media for recovery of mycobacteria in clinical specimens. *Journal of Clinical Microbiology*, *37*(5), 1366–1369. https://doi.org/10.1128/JCM.37.5.1366-1369.1999.

Song, W., Zhao, J., Zhang, Q., Liu, S., Zhu, X., An, Q., Xu, T., Li, S., Liu, J., Tao, N., Liu, Y., Li, Y., & Li, H. (2021). COVID-19 and tuberculosis coinfection: An overview of case reports/case series and meta-analysis. *Frontiers in Medicine*, *8*. www.frontiersin.org/articles/10.3389/fmed.2021.657006.

Spence, D. P., Hotchkiss, J., Williams, C. S., & Davies, P. D. (1993). Tuberculosis and poverty. *BMJ: British Medical Journal*, *307*(6907), 759–761.

Steingart, K. R., Dendukuri, N., Henry, M., Schiller, I., Nahid, P., Hopewell, P. C., Ramsay, A., Pai, M., & Laal, S. (2009). Performance of purified antigens for serodiagnosis of pulmonary tuberculosis: A meta-analysis. *Clinical and Vaccine Immunology: CVI*, *16*(2), 260–276. https://doi.org/10.1128/CVI.00355-08.

Stochino, C., Villa, S., Zucchi, P., Parravicini, P., Gori, A., & Raviglione, M. C. (2020). Clinical characteristics of COVID-19 and active tuberculosis co-infection in an Italian reference hospital. *European Respiratory Journal*, *56*(1). https://doi.org/10.1183/13993003.01708-2020.

Sumner, A., Hoy, C., & Ortiz-Juarez, E. (2020). *Estimates of the impact of COVID-19 on global poverty* (Working Paper No. 2020/43). WIDER Working Paper. https://doi.org/10.35188/UNU-WIDER/2020/800-9.

Tadolini, M., Codecasa, L. R., García-García, J.-M., Blanc, F.-X., Borisov, S., Alffenaar, J.-W., Andréjak, C., Bachez, P., Bart, P.-A., Belilovski, E., Cardoso-Landivar, J., Centis, R., D'Ambrosio, L., Souza-Galvão, M.-L. D., Dominguez-Castellano, A., Dourmane, S., Jachym, M. F., Froissart, A., Giacomet, V., . . . Migliori, G. B. (2020). Active tuberculosis, sequelae and COVID-19 co-infection: First cohort of 49 cases. *European Respiratory Journal*, *56*(1). https://doi.org/10.1183/13993003.01398-2020.

Tadolini, M., García-García, J.-M., Blanc, F.-X., Borisov, S., Goletti, D., Motta, I., Codecasa, L. R., Tiberi, S., Sotgiu, G., & Migliori, G. B. (2020). On tuberculosis and COVID-19 co-infection. *The European Respiratory Journal*, *56*(2), 2002328. https://doi.org/10.1183/13993003.02328-2020.

Uplekar, M., Weil, D., Lonnroth, K., Jaramillo, E., Lienhardt, C., Dias, H. M., Falzon, D., Floyd, K., Gargioni, G., Getahun, H., Gilpin, C., Glaziou, P., Grzemska, M., Mirzayev, F., Nakatani, H., Raviglione, M., & for WHO's Global TB Programme. (2015). WHO's new end TB strategy. *Lancet (London, England)*, *385*(9979), 1799–1801. https://doi.org/10.1016/S0140-6736(15)60570-0.

Vynnycky, E., & Fine, P. E. (1999). Interpreting the decline in tuberculosis: The role of secular trends in effective contact. *International Journal of Epidemiology*, *28*(2), 327–334. https://doi.org/10.1093/ije/28.2.327.

Wang, D., Hu, B., Hu, C., Zhu, F., Liu, X., Zhang, J., Wang, B., Xiang, H., Cheng, Z., Xiong, Y., Zhao, Y., Li, Y., Wang, X., & Peng, Z. (2020). Clinical characteristics of 138 hospitalized patients with 2019 novel coronavirus–infected pneumonia in Wuhan, China. *JAMA*, *323*(11), 1061–1069. https://doi.org/10.1001/jama.2020.1585.

Wang, M., Cao, R., Zhang, L., Yang, X., Liu, J., Xu, M., Shi, Z., Hu, Z., Zhong, W., & Xiao, G. (2020). Remdesivir and chloroquine effectively inhibit the recently emerged novel coronavirus (2019-nCoV) in vitro. *Cell Research*, *30*(3), 269–271. https://doi.org/10.1038/s41422-020-0282-0.

Wang, Z., Yang, B., Li, Q., Wen, L., & Zhang, R. (2020). Clinical features of 69 cases with coronavirus disease 2019 in Wuhan, China. *Clinical Infectious Diseases*, *71*(15), 769–777. https://doi.org/10.1093/cid/ciaa272.

Whittier, S., Hopfer, R. L., Knowles, M. R., & Gilligan, P. H. (1993). Improved recovery of mycobacteria from respiratory secretions of patients with cystic fibrosis. *Journal of Clinical Microbiology*, *31*(4), 861–864.

World Health Organization. (2013). Automated real-time nucleic acid amplification technology for rapid and simultaneous detection of tuberculosis and rifampicin resistance: Xpert MTB (No. WHO/HTM/TB/2013.16). World Health Organization.

World Health Organization. (2014). Xpert MTB/RIF implementation manual: Technical and operational 'how-to'; Practical considerations (No. WHO/HTM/TB/2014.1). World Health Organization.

World Health Organization. (2020). WHO Director-General's remarks at the media briefing on 2019-nCoV on 11 February 2020.

Xu, Z., Shi, L., Wang, Y., Zhang, J., Huang, L., Zhang, C., Liu, S., Zhao, P., Liu, H., Zhu, L., Tai, Y., Bai, C., Gao, T., Song, J., Xia, P., Dong, J., Zhao, J., & Wang, F.-S. (2020). Pathological findings of COVID-19 associated with acute respiratory distress syndrome. *The Lancet Respiratory Medicine*, *8*(4), 420–422. https://doi.org/10.1016/S2213-2600(20)30076-X.

Yasri, S., & Wiwanitkit, V. (2020). Tuberculosis and novel Wuhan coronavirus infection: Pathological interrelationship. *Indian Journal of Tuberculosis*, *67*(2), 264. https://doi.org/10.1016/j.ijtb.2020.02.004.

5 Coinfection of SARS-CoV-2 with Viruses Causing Respiratory and Systemic Infections

Sabiha Fatima, Samina Wasi, Nikhat J. Siddiqi and Reem Hamoud Alrashoudi

5.1 INTRODUCTION

Coronavirus disease 2019 (COVID-19) was first reported in Wuhan, China, in December 2019. Due to its rapid spread to several other countries, on 11th March 2020, the World Health Organization (WHO) declared COVID-19 as a pandemic. The *International Committee for Taxonomy of Viruses* classified the causative agent of COVID-19 as "Severe Acute Respiratory Syndrome Coronavirus-2 (SARS-CoV-2)". This is because the genome sequence analysis of SARS-CoV-2 showed 79% sequence identity with SARS-CoV that instigated the outbreak in 2003, which also originated in China (Chen et al., 2020b, Lu et al., 2020). By April 2021, the SARS-CoV-2 virus had spread to over 223 countries and territories infecting 137 million people (Li et al., 2022). According to the most recent data, 542.18 million confirmed cases of COVID-19 and 6,329,375 associated deaths were documented by June 28, 2022 (WHO, 2022).

The COVID-19 pandemic affected different countries with variable severity. A meta-analysis review reported lowest prevalence of viral co-infection with COVID-19 in Saudi Arabia and the highest in China (Malekifar et al., 2021). It has been established before the COVID-19 pandemic that co-infections of the respiratory tract with viruses significantly increase the likelihood of hospital admission as compared to infections by single virus (Pinky and Dobrovolny, 2016). During the pandemic, the clinical symptoms and treatment outcomes differed in SARS-CoV-2 patients who were and were not coinfected (Alhumaid et al., 2021). This may be one of the reasons for China to witness a high COVID-19 related morbidity rate, whereas it remained moderate in Saudi Arabia. However, this hypothesis cannot be proposed with certainty due to lack of documented patient records and history. Moreover, lack of adequate diagnostic equipment further

DOI: 10.1201/9781003324911-7

pose challenges in accurate assessment of SARS-CoV-2 co-infections with other viruses. However, the overall understanding is suggestive of higher prevalence rates of coinfecting viruses than reported in literature. Hence, relevant documentation and meta-analysis remains the gold standard in analysis of viral co-infections with SARS-CoV-2 and can lead to improved diagnosis and treatment outcomes (Shen et al., 2020).

Clinical manifestations of COVID-19 range from asymptomatic infections to pneumonia induced acute respiratory distress syndrome (CARDS), systemic dysfunctions such as sepsis, septic shock and multi-organ dysfunction or failure (Torres Acosta and Singer, 2020; Vos et al., 2021). Like typical respiratory infections, COVID-19 is transmitted primarily by droplets released during coughing or sneezing (Chen et al., 2020a). Based on our current understanding, the SARS-CoV-2 affects the upper respiratory tract, causing typical symptoms such as fever, cough and shortness of breath, though the severity of the infection is inversely proportional to the host immune response. Some patients experience acute symptoms such as nausea, exhaustion, headaches and mild fever, on infection with SARS-CoV-2 (Donyavi et al., 2021). However, it often results in pneumonia among those exposed to high viral load (Chu et al., 2020). In COVID-19, age is identified as an important factor that affects the rapidity of disease progression. Additionally, comorbidities including hypertension, coronary heart disease, chronic obstructive pulmonary disease, diabetes mellitus, malignancies and HIV infection lead to life-threatening conditions in patients (Ejaz et al., 2020).

The co- and mixed-infections exhibit extended mechanisms of inhibiting the host immune response. Co-infections typically occur within one to four days of COVID-19 disease onset. Among viral co-infections, the rate of systemic infections has been reported to be three times higher than respiratory infections in COVID-19 patients (Malekifar et al., 2021). A higher rate of systemic co-infections as compared to respiratory co-infections in COVID-19 is also reported in other studies (Lansbury et al., 2020; Musuuza et al., 2021; Krumbein et al., 2021). The exact mechanism of co-infections with SARS-CoV-2 in triggering immune responses remains unclear, though the lowered drug response or drug resistance in such cases is well reported (Russell et al., 2021; Lucien et al., 2021; Knight et al., 2021). Thus, co-infections in COVID-19 exacerbate the progression of the disease, exhibit lower treatment and prophylactic responses and increase morbidity and mortality rates during pandemics.

Presently, the interaction of SARS-CoV-2 with different viruses is thoroughly investigated to gain relevant insights into the clinical outcomes of co-infections. The present chapter provides an overview of the common systemic and respiratory co-infections that occur along with SARS-CoV-2.

5.2 RESPIRATORY VIRAL CO-INFECTION IN COVID-19

A variety of co-circulating viruses within the human respiratory tract may cause acute respiratory tract infections when our immunity is compromised (Babiuk et al., 1988). Mixed infections, however, substantially impact respiratory functions (Kiseleva et al., 2020). During outbreak of epidemics that target the respiratory system, it is possible for respiratory viruses to act as opportunistic pathogens and cause co-infection. In fact, it is estimated that over 50% of deaths related to COVID-19 may be caused by mixed infections (Lai et al., 2020).

Typically, the viruses attack the host's immune system. While doing so, the viruses causing respiratory infections weaken the defence mechanisms in and around the lungs, increasing its susceptibility to opportunistic bacterial and viral infections and causing secondary pneumonia (Mirzaei et al., 2020). In COVID-19, co-infection of respiratory viruses was estimated to occur in ~1.4% of cases, which was lower compared to bacterial and fungal co-infection, in a study conducted in a tertiary care hospital in Singapore (Wee et al., 2020). Though the reports based on meta-analysis and systemic reviews of global data suggest the higher co-infection rate of SARS-CoV-2 with respiratory viruses (Lansbury et al., 2020; Davis et al., 2020; Alhumaid et al., 2021).

Several reports indicate co-infection of SARS-CoV-2 with respiratory viruses and a rapid progression to severe COVID-19 disease conditions in such instances. A case study from Wuhan, China reported that 5.8% of the confirmed COVID-19 patients were coinfected with other respiratory viruses (Peci et al., 2021). Besides, co-infection with fungi and bacteria accounts for the second most common occurrence of COVID-19 in China (Zhu et al., 2020a). Another case study from Northern California also reported a similar trend where co-infection with viruses was more common among 20.7% of COVID-19 patients with mixed infections (Kim et al., 2020).

The commonly reported viruses that co-infect with SARS-CoV-2 are respiratory syncytial virus (RSV), influenza A and B, parainfluenza, enterovirus, metapneumovirus, hepatitis B (HBV), dengue, Epstein-Barr (EBV), cytomegaloviruses (CMV), human immunodeficiency virus (HIV) and non-SARS-CoV-2 coronaviruses (Aghbash et al., 2021).

Among these, the enterovirus, non-SARS-CoV-2 coronaviruses, RSV and influenza A virus co-infections are distinctly reported in several COVID-19 studies (Musuuza et al., 2021). In fact, one of the earliest case studies that reported co-infection in COVID-19 described four patients who were co-infected with influenza virus (Cuadrado-Payán et al., 2020). In a recent study published in Lancet, 3.5% cases of co-infection were reported among 212,466 patients, and 8.4% of co-infection occurred with influenza, RSV or adenovirus (Swets et al., 2022).

Very few studies have attempted to understand the mechanisms involved in viral co-infections with SARS-CoV-2. From these studies, a correlation of enhanced inflammatory marker with increased serum levels of tumour necrosis factor-α (TNF-α) and interleukins (IL-6 and IL-10), with a weakened immune system, can be easily comprehended. This is coupled with a significant decline in the lymphocyte count (< 800) and high serum levels of lactate dehydrogenase (LDH >350 U/L), ferritin and troponin (>1000 ng/ml) and D-dimer (>1 mcg/ml) in patients (Samprathi and Jayashree, 2021; Zhu et al., 2020a). The key characteristics of common viral co-infections with SARS-CoV-2 are described in this section.

5.2.1 Co-Infection of SARS-CoV-2 with Influenza Virus

As indicated earlier in this section, influenza is among the commonly reported co-infections with SARS-CoV-2. Studies from China reported that co-infection of SARS-CoV-2 and influenza virus was common during the initial outbreak of COVID-19 in Wuhan, and patients with co-infection faced a higher risk of poor health outcomes (Yue et al., 2020; Cuadrado-Payán et al., 2020). Sakamoto et al. (2020) suggested that the increased co-infection rate may be due to the common prevalence of influenza virus globally. They further presented more convincing evidence of undiagnosed cases of influenza co-infections with SARS-CoV-2 based on epidemiological data of influenza infections between 2014 and 2020, where they indicated influenza incidences to be lower in 2020 compared to previous years. Their debate is reasonably rational and is further supported by Dadashi et al. (2021), who reported ~1.2% co-infection rates based on meta-analysis of published data.

The SARS-CoV-2 and influenza are presented with similar clinical symptoms like cough, headache, high-grade fever and pneumonia (Ding et al., 2020). The mode of transmission and seasonal (cold) preference for infection is also similar (John et al., 2020). Moreover, SARS-CoV-2 predominantly infects the alveolar type II pneumocytes which are also the major sites of influenza virus replication (Bai et al., 2022). The SARS-CoV-2 co-infection with influenza A virus is associated with more severe complications, especially in high-risk individuals and the elderly patients, as compared to co-infection with influenza B virus. The co-infection with influenza A aggravates the symptoms and also complicates the diagnosis and treatment of patients (Lansbury et al., 2020; Yue et al., 2020). Besides, the underlying diseases, such as heart disease, asthma, chronic neurological pathologies, diabetes and diabetes-related retinopathy may contribute to increased mortality in patients with co-infection (Hashemi et al., 2021).

A recent study confirmed that influenza virus re-infections promote entry and infectivity of SARS-CoV-2 virus in cells, thus contributing to the severity of COVID-19 (Bai et al., 2022). On a molecular level, the co-infection with influenza augments neutrophil activation, leading to an excessive cytokine production. In turn, the extreme levels of cytokines trigger massive infiltration of neutrophils and macrophages into the infected region which results in shock, myocarditis, acute respiratory distress syndrome (ARDS) and organ dysfunction, including acute kidney injury and serious pulmonary sequelae (Kulkarni et al., 2019, Cao et al., 2020; Ma et al., 2020; Wu et al., 2020). The disastrous outcome of SARS-CoV-2 co-infection with influenza A virus and comparatively less severity of SARS-CoV-2 infection is also confirmed in animal (hamster) models (Zhang et al., 2021).

Studies published in *Lancet* have reported more cases of patients requiring invasive mechanical ventilation in co-infections than SARS-CoV-2 infections (Cuadrado-Payán et al., 2020; Swets et al., 2022). Considering the above risks of SARS-CoV-2 co-infection with influenza A virus, a contrary finding of lower risk of mortality was reported in China (Guan et al., 2021). This can possibly be explained with the existence of varied influenza strains among different countries as well as the herd immunity of a population against these strains.

Although, the clinical manifestations of SARS-CoV-2 and influenza infections are characterised with similar patterns, they do differ in intensity of triggering host immune response. Influenza is mostly asymptomatic or causes moderate illness, whereas SARS-CoV-2 infection develops complications for five to seven days, which in many cases may lead to serious illness (Jiang et al., 2020). In influenza, the virus is shed within five to ten days of infection, while it takes two to five weeks in SARS-CoV-2 infections. Also, the prevalence of ARDS is rare in influenza outbreaks (< 1% mortality rate) as compared to COVID-19 (3–4% mortality rate) (Fang et al., 2020; Jin et al., 2020).

Overall, it can be suggested that co-infection of SARS-CoV-2 and influenza potentiates chronic illness and eventually disrupts the patient's response to therapeutic agents, thus necessitating simultaneous diagnosis of these viruses for effective care and treatment (Ji et al., 2020).

5.2.2 Co-Infection of SARS-CoV-2 with Human Coronavirus (CoVs)

The human coronaviruses (HCoVs) are single-stranded RNA viruses with a relatively large genome size which makes it highly prone to mutations. A recombination rate of over 25% is reported throughout the entire viral genome. Of the seven distinguished HCoVs, three strains (MERS-CoV, SARS-CoV and SARS-CoV-2) are associated with epidemic outbreaks. The other four strains (HCoV-NL63, HCoV-OC43, HCoV-229E and HCoV-HKU1) are endemic which has prevailed among humans for decades. They commonly cause co- or mixed-infections with each other and other respiratory viruses. Characteristically, they develop mild symptoms such as common cold and lower respiratory tract infections in most infected individuals that are resolved within one to two weeks without medication. However, co-infection of endemic strains with MERS-CoV, SARS-CoV and SARS-CoV-2 significantly accentuates the patients' clinical condition. To date, co-infections of SARS-CoV and MERS-CoV are reported in multiple cases, whereas co-infection of one of these viruses with SARS-CoV-2 is relatively rare (Lai et al., 2020). However, these rare instances of SARS-CoV-2 and SARS-CoV co-infection are believed to trigger a cascade of inflammatory responses that cause lung damage (Tay et al., 2020; Aghbash et al., 2021). A recent study has reported the co-infection of SARS-CoV-2 variants, that is Omicron and Delta, in two patients with chronic kidney disease who showed a high risk of mortality (Rockett et al., 2022). Both patients required frequent dialysis and had a severely compromised immunity. In this case, it was suggested that, most likely, the sustained exposure of these patients to an infectious zone (due to frequent hospital visits) led to this co-infection

A case study from Hong Kong reported fatal consequence of SARS-CoV-2 co-infection with HCoV-229E (Lau et al., 2021). Since HCoV-229E causes minor symptoms on infection, the standard laboratory panels rarely devise standards for detection of respiratory viruses. This has created a major challenge in data analysis of co-infection rates. It is apparent that the studies analysing co-infection in patient samples simply underestimated the chance factor of HCoV-229E or other endemic HCoV strains or lacked standard diagnostic references for their detection.

Contrary to the fatal consequences and aggravated immune response in most co-infections, Dugas et al. (2021) confirmed cross-reactivity between SARS-CoV-2 and HCoV-OC43 on co-infection. They also suggested that anti-OC43 antibodies can protect patients against the severity of COVID-19. Guo and his colleagues further demonstrated that S-IgG antibodies to HCV-OC43 are increased in patients with SARS-CoV-2, and the level of expression correlates with the severity of the disease (Guo et al., 2021). However, relevant insights could not be deduced from their study since none of the COVID-19 patients were co-infected with HCV-OC43. Although supporting data is unavailable at present, the above co-infection rate of only 0.08% in children with respiratory symptoms was indicated in another study.

In Canada, a study analysed 298,415 respiratory specimens of COVID-19 symptomatic patients and reported 18 co-circulating pathogens which included all four endemic HCoVs (Marshall et al., 2021). They further reported very low rates of SARS-CoV-2 co-infection as compared to non-SARS-CoV-2 co-infection. A critical diagnostic insight was provided by a case from Singapore that reported co-infection of HCoV-HKU1 with SARS-CoV-2 in a patient who was tested as false negative for COVID-19 three times (Chaung et al., 2020). Overall, these data suggests that co-infections between HCoVs are common, and a high degree of suspicion is required to make the diagnosis. More than one repetition should be proposed at appropriate intervals to ensure a negative COVID-19 test. More importantly, the co-infections of SARS-CoV-2 and other HCoVs among susceptible children should not be ignored.

5.2.3 Co-Infection of SARS-CoV-2 with Adenovirus

Although less frequently, the co-infection of SARS-CoV-2 and adenoviruses is reported in literature. A recent study investigated over 200,000 COVID-19 specimens for signs of co-infection. They reported only 135 specimens of SARS-CoV-2 and adenovirus co-infections among ~7000 co-infected samples. All these patients presented a higher possibility of an adverse clinical outcome. However, unlike influenza virus co-infection, adenovirus co-infection was not associated with the necessity of invasive mechanical ventilation (Swets et al., 2022). Motta and Gómez (2020) reported that adenovirus co-infection along with poorly controlled diabetics in COVID-19 leads to elevated levels of D-dimer, lactate dehydrogenase and ferritin. It further progresses into a life-threatening condition due to hypoxia and ARDS. The adenovirus and SARS-CoV-2 co-infection are more commonly reported in patients over 15 years of age; though few data also suggests mixed infections (including adenovirus and SARS-CoV-2) in children less than 10 years of age (Motta and Gómez, 2020).

Overall, several factors like increased susceptibility amongst young children to co-infection and possibility of novel adenovirus strains remain unclear due to less prevalence of SARS-CoV-2 and adenovirus co-infection. Nevertheless, it is better to rule out the possibility of co-infection with treatable pathogens.

5.3 SYSTEMIC VIRAL CO-INFECTION IN COVID-19

The COVID-19 disease is often complicated by concurrent systemic viral infections caused by HIV, CMV, EBV and HBV. It is suggested that co-infection of SARS-CoV-2 with these viruses may have caused the COVID-19 pandemic (Blanco et al., 2020, Simonnet et al., 2021). In contrast to respiratory viruses, blood-borne viruses were reported to have the highest rate of co-infection in COVID-19 (Malekifar et al., 2021; Lee et al., 2021). Thus, the diagnosis of co-infecting blood-borne viruses is crucial in the treatment of COVID-19 disease.

On a molecular level, the activation of cytokines (including chemokine) related cascades, macrophages and NK cells have been recognised as key players in progression of the severity of COVID-19 infection. They control the spread of viral infections by lysing the infected cells (Ardolino et al., 2014; Angka et al., 2018). However, in COVID-19 cases, elevated levels of cytokines and chemokines (TNF-α, IL-6, CXCL10, CCL2, CCL3, CCL5, IL-6, IL-10 and TGF-β) down-regulate NKG2D and suppress the antiviral activity of NK cells (Mazzoni et al., 2020; Zheng et al., 2020). Simultaneously, they also cause dysfunction of cytotoxic T cells and inactivate CD4+ immune cells (Mehta et al., 2020). Additionally, the CXCR3 pathway is activated in COVID-19 patients that recruit NK cells from the peripheral blood to the lungs (Lagunas-Rangel and Chávez-Valencia, 2020). Consequently, the lung tissues are infiltrated with cytokines secreted by NK cells, and their concentration is reduced in the peripheral blood (leading to compromised immunity). There is further evidence of hyperactive lung macrophages to cause a surge of cytokine infiltration in lung tissues in COVID-19. This overactivation of immune response is commonly referred to as 'cytokine storms' and results in immune exhaustion (Ghasemzadeh et al., 2022). The level of immunity exhaustion markers like LAG3, HAVCR2, NKG2A and CD94 are further reported to be significantly elevated during co-infection of SARS-CoV-2 with respiratory or systemic viral infections (Yaqinuddin and Kashir, 2020; Zheng et al., 2020).

It is not surprising that the systemic viral infections that trigger cytokine storms (like CMV, EBV and HBV) lead to severe clinical manifestations and poor therapeutic outcome on co-infection with SARS-CoV-2 (Blanco et al., 2020; Simonnet et al., 2021; Wilk et al., 2020). Also, during co-infections, factors like SOCS1 and SOCS3 mutations, cryptic antigen expression, and epitope spreading are reported to cause a common complication of COVID-19 known as immune thrombocytopenia (platelet count less than 100 x 10^9/L). It is more commonly reported in SARS-CoV-2 co-infection with HBV, HCV, CMV, varicella-zoster virus (VZV), HIV and zika virus (Bhattacharjee and Banerjee, 2020; Pascolini et al., 2021). In this section, we report the characteristics of commonly reported systemic co-infections in COVID-19.

5.3.1 SARS-CoV-2 and HIV Co-Infections

The HIV co-infection shows classic symptoms due to immunodeficiency and slower antibody formation. Hence, the associated increase in susceptibility to SARS-CoV-2 and HIV/related infections occurs easily, and the resulting outcome is fatal in over 20% of cases (Zhu et al., 2020b; Dadashi et al., 2022; Danwang et al., 2022). A low CD4+ T cell count in HIV-positive patients and a surge of cytokines on SARS-CoV-2 infection are linked to severe organ damage during co-infection (Danwang et al., 2022). Several studies have suggested that the viral load of HIV in co-infection is proportional to the severity of COVID-19 (Zhu et al., 2020b; Vizcarra et al., 2020). In contrast, it is also suggested that HIV co-infection does not necessarily increase the severity of COVID-19 disease; though SARS-CoV-2 infection in later stages among HIV-positive individuals may speed the progression towards AIDS (Hariyanto et al., 2021; Falutz et al., 2021). This condition can worsen in elderly individuals and those suffering from comorbidities (Tang et al., 2020; Danwang et al., 2022).

The clinical symptoms of SARS-CoV-2 and HIV co-infection are rarely reported in literature. One of the reasons may be the high fatality associated with this co-infection (Danwang et al., 2022). However, the above reasoning cannot be proclaimed with assertion due to rare documentation of characteristics of disease progression and death rate in COVID-19 co-infection with HIV. It is highly possible that the conditions like high blood pressure, obesity, diabetes and chronic kidney/liver disease along with HIV co-infections in COVID-19 severely contribute to immune dysfunction and persistence of mixed infections (Zhao et al., 2020a; Qin et al., 2020). Thus, the vulnerability to SARS-CoV-2 co-infection in HIV-positive patients can be related to a compromised immunity.

More disturbingly, an early case study of a HIV co-infected COVID-19 patient from China reported high degree of drug resistance of SARS-CoV-2 towards reverse transcriptase nucleoside inhibitors, non-nucleoside reverse transcriptase inhibitors as well as their combinations (Livingston and Bucher, 2020). An increased D-dimer level along with destruction of fibrinogens, which leads to cyanosis, ischemia and ultimately death, is another disturbing characteristic of SARS-CoV-2 and HIV co-infection reported by several authors (Léonard-Lorant et al., 2020; Shah et al., 2020; Levi et al., 2020). In addition, the poor outcomes of HIV co-infection in COVID-19 are not only related to increase in severity of disease but also the highly variable characteristic observations and treatment outcomes reported in such patients (Geretti et al., 2021; Nomah et al., 2022). The infectivity of different HIV strains, their impact on immune system and delayed antibody response is also poorly understood. This creates another big hurdle in understanding the immune response and disease prognosis during co-infection of HIV with any other opportunistic pathogens. In addition, various factors like impact of lymphopenia, thrombocytopenia and distribution of immune cells during co-infection is challenging to elucidate based on limited case studies (Aghbash et al., 2021).

Overall, the lack of observational data on relevance of age, comorbidities and HIV strains contribute to confounding factors in analysis of SARS-CoV-2 and HIV co-infection and remain highly controversial.

5.3.2 SARS-CoV-2 and CMV Co-Infections

The CMV has been recognised as a major threat to immune risk profile among elderly individuals and leads to fatal outcome within two years in spite of critical management of infection (Wikby et al., 2002). The immunosenescence and immunosuppression associated with CMV increases the death risk on co-infection with other pathogens including SARS-CoV-2 (Merani et al., 2017). More commonly, the CMV is identified as a latent virus which is activated by other viral co-infections. In COVID-19, lymphocytopenia leads to cellular immune system deficiencies which, in turn, activate the CMV (D'Ardes et al., 2020).

CMV infection triggers production of IL-6, whereas increased levels of IL-6, IL-10 and TNF-α are reported in COVID-19 patients. Pathologically, the CMV co-infection in COVID-19 is suggested to aggravate the disruption of peripheral blood T-cells differentiation (leading to $CD4^+$ and $CD8^+$ dysfunction) and up-regulate cytokine production (Zheng et al., 2020). Angiotensin-converting enzyme 2 (ACE2; receptor in vital organs) is identified as the major entry portal for SARS-CoV-2 (Jackson et al., 2022). In instances of infection/activation of CMV in COVID-19 patients, it activates the cytokine storm. The resulting surge of cytokines infiltrated at infection site during co-infection, along with diminished IFN-γ secretion, almost entirely compromises the immune response in patients (Lippi et al., 2020). In COVID-19 patients, particularly the CMV-driven pneumonitis is identified as the key driver of lung function compromise (Naendrup et al., 2021; Niitsu et al., 2021).

To make matters worse, the anti-cytokine therapies and administration of immune modulators are associated with serious pulmonary complications in SARS-CoV-2 and CMV co-infection (Antinori et al., 2020; Kimmig et al., 2020). Distinct studies of SARS-CoV-2 and CMV co-infection among kidney transplant recipients have been reported in literature; all of which suggest higher rates of organ failure (Banerjee et al., 2020; Billah et al., 2020). These studies reported common findings of increased levels of creatine kinase and blood urea nitrogen and associated it with cytokine storms triggered by both CMV as well as SARS-CoV-2 infections.

A discouraging finding was reported in a case study of CMV reactivation following SARS-CoV-2 vaccination. The vaccine induced reactivation of CMV was characterised by a paradoxical worsening of symptoms and led to development of COVID-19 with intense inflammatory response. The severity was further aggravated on administration of antiretroviral therapy (Plüß et al., 2022). The ineffectiveness of antiretroviral therapies in vaccine induced co-infections is also reported in other studies (Nicoli et al., 2022; Gushchin et al., 2022). In another study Weber et al. (2022), reported a very high prevalence of CMV seropositivity among non-geriatric patients (< 60 years) with severe COVID-19. Considering both scenarios, the very practical approach of controlling COVID-19 pandemic through vaccinations, may in fact increase its spread. At the same time this can increase the severity and drug resistance due to increased risk of co-infections.

Overall, CMV activation is a majorly underestimated threat on COVID-19 vaccination, and urgent measures are required for critical awareness of this scenario. Besides, CMV co-infection in COVID-19 presents a classic example of cytokine storms and compromised immunity in patients leading to organ failure, coagulopathy and other complications reported in literature.

5.3.3 SARS-CoV-2 and EBV Co-Infections

The EBV is associated with infectious mononucleosis that leads to lymphomas and carcinomas. Since this condition is aggravated by compromised immune system, SARS-CoV-2 and EBV co-infection result in fatal outcome by progression through dysplasia and systemic malignancy (Nadeem et al., 2021).

Generally, the EBV infections are manifested in the form of symptoms like fever, pharyngotonsillitis, cervical lymphadenopathy and hepatosplenomegaly. Additionally, the peripheral blood samples show an increase in atypical lymphocytes. An unusual case study reported these characteristic symptoms in a young woman (26 years old) who tested negative for anti-Epstein-Barr virus immunogens (Fukuda et al., 2022). Subsequently, she tested positive for SARS-CoV-2 infection. Also, other common infections prevalent during the pandemic were ruled out. Based on this finding, the cross-reactivity of EBV with SARS-CoV-2 was suggested, which resulted in elimination of EBV (Chen et al., 2021). Supporting evidence can also be drawn from reviews published by Ju et al. (2020) and Reche (2020), which suggested that the cross-reactive epitopes of EBV (APSASAPF and DLLLDASVEI) may be drivers of protective immune responses in individuals exposed to SARS-CoV-2.

In contrast, 55.2% of COVID-19 patients who were co-infected with EBV showed signs of impaired cellular immune function, inflammation, stress, anxiety and depression (Chen et al., 2021). A significant correlation is also reported between $CD8^{+}$T cells and EBV DNA levels and COVID-19 severity in several studies (Liu et al., 2020; Paolucci et al., 2021). Relatively higher levels of AST, ALT and C-reactive proteins are also reported in EBV/SARS-CoV-2 co-infections along with high-grade fever (Xiong et al., 2020; Roncati et al., 2020).

Consequently, instances of liver damage is commonly reported in EBV/SARS-CoV-2 co-infections with elevated levels of total bilirubin and low albumin in ~50% cases, as suggested in systematic reviews of global data (Meng et al., 2022; Zubchenko et al., 2022) These studies further reported progression of COVID-19 severity with complications like multiple supra-/subdiaphragmatic lymphadenopathies, interstitial pneumonia and axillary adenomegaly. Reactivation of EBV is another common possibility in immunocompetent COVID-19 patients. The co-infections have been linked to cholestatic hepatitis (Da Cunha et al., 2022), lymphoproliferative disease (Roncati et al., 2020) and ganciclovir treatment (Meng et al., 2022) among others. These case studies reported high mortality associated with EBV reactivation and co-infection with COVID-19.

Overall it can be summarised that EBV/SARS-CoV-2 co-infection in patients causes most unusual complications that may challenge the diagnosis of the disease. Additionally, although the cross-reactivity of EBV is well established with other viral pathogens, its interaction with SARS-CoV-2 is poorly understood due to limitations of information related to sample size, time and severity of EBV infections in published literature.

5.3.4 SARS-CoV-2 and HBV Co-Infections

Research indicates that the HBV virus can be suppressed or reactivated following co-infection with another virus. Owing to the similar transmission modes of HIV and HBV, their co-infection is a common occurrence. In this case the compromised immune system due to HIV leads to development of chronic hepatitis B (Hoffmann and Thio, 2007). A compromised immunity is also reported in severe COVID-19 cases. Probably this leads to progression of viral hepatitis into complex case of liver dysfunction and increased mortality rate on exposure to SARS-CoV-2 (Moon et al., 2020). Although the above reasoning is suitably backed up by several reports, many conflicting case studies challenge our current understanding of SARS-CoV2 and HBV co-infections. For instance, it is reported that secretion of cytokines such as IL-6 and TNF-α suppress replication and pathogenicity of HBV (Xia and Protzer, 2017). Given the common pathogenesis of COVID-19 resulting in cytokine storms, suppression of HBV is also reported on co-infection with SARS-CoV-2 or HCV (Han et al., 2020; Zhao et al., 2020b; Cheng et al., 2020).

HBVs are among the globally prevalent viruses infecting over 300 million people globally (Wu et al., 2021). The scenario of suppression of HBV can be argued based on epidemiological data that indicates a continuous rise in HBV infections with or without SARS-CoV-2 co-infection during the COVID-19 pandemic (Yang et al., 2022). In this context, a higher possibility of liver dysfunction and life-threatening complications are reported due to SARS-CoV-2 and HBV co-infection. Additionally, the mortality rate of SARS-CoV-2 and HBV co-infection is reported to be 10.5% higher than those with SARS-CoV-2 infection (Lin et al., 2021). The severity of disease was also reported to be higher during co-infection. Most importantly, considering the impact of both infections on liver function, global prevalence of both pathogens and persistence of their co-infection for long time, it is highly suggestive of detrimental disease outcome on co-infections (Kariyawasam et al., 2022; Marjot et al., 2021).

The progression of HBV co-infection to severely complicated COVID-19 disease is commonly reported in literature with common indications of liver dysfunction (with elevated levels of AST and ALT) and simultaneous cytokine storm. This key characteristic is reported in nearly all reports that have analysed the co-infection characteristics of HBV and SARS-CoV-2. Also, the primary outcome of most cohort studies is the dead patients with or without long-term admission to the ICU (Kariyawasam et al., 2022; Yang et al., 2020; Jothimani et al., 2020).

Overall, patients with SARS-CoV-2 and HBV co-infection suffer risks of poor prognosis and extremely high rates of liver failure, multi-organ damage and death. Hence, diagnosis of HBV co-infection may be prescribed as standard in COVID-19 testing since critical and advanced care of patients may improve the life expectancy and treatment outcome.

5.4 VIRAL CO-INFECTIONS: SYMPTOMS AND DIAGNOSIS

Aghbash et al. (2021) describes the clinical signs and symptoms associated with COVID-19. Common symptoms like fever and cough is reported in over 70% of cases followed by shortness of breath, myalgia and sore throat, which is experienced in approximately 27.7% cases. In less than 7.5% cases, individuals (most likely with a high immunity) have been reported to experience a slight headache, diarrhoea or rhinorrhea in COVID-19. Unusual symptoms like anosmia, diarrhoea and throat pain are also documented in association with COVID-19 (Feng et al., 2020).

Increase in intensity of above symptoms is widely reported in COVID-19 co-infections (Ding et al., 2020). One of the major reasons for misdiagnosis of SARS-CoV-2 infection by radiological imaging is the associated difficulty in differential diagnosis of other viral respiratory tract infections. For this reason, RT-PCR is the most reliable method for diagnosis of SARS-CoV-2 and other viruses. It is a highly sensitive method; however, factors like sampling techniques, time of sampling and accuracy in programming of RT-PCR techniques raises a risk of bias and false-negative reports (Lansbury et al., 2020). Hence, as indicated earlier in this chapter, the necessity to repeat the test should be critically considered in severely ill patients (Chaung et al., 2020).

The co-infections are reported to be persistent and highly transmissible. They also cause rapid progression of clinical symptoms associated with either infection. In turn, these factors significantly challenge the prognosis and management of co-infections (Gushchin et al., 2022; Weber et al., 2022). The co-infections are also suggested to cause elevation in levels of LDH, ALT, AST, D-dimer, troponin and CRP along with lymphopenia, prolonged prothrombin time, neutrophilia and eosinophilia (Harrison et al., 2020). The most common complications of COVID-19 co-infections are respiratory failure, dysphagia, seizure, organ failure, neurological disorders and low drug response (Hashemi et al., 2021; Da Cunha et al., 2022). Although diagnosis of co-infecting pathogens is challenging, due to the opportunistic nature of many pathogens, the increased intensity of symptoms should be considered as possible mixed infection in COVID-19. Having said that, it is necessary to rule out any infection based on differential diagnosis (liver diseases and hepatitis infection) and those contributing to compromised immunity (HIV and related infections) in patients (Lansbury et al., 2020; Chaung et al., 2020). Additionally, special diagnostic panels can be devised based on infections that are endemic to a particular region to improve the probability of detecting co-infections and reducing the associated complications (Dugas et al., 2021; Lau et al., 2021). Collectively these steps may help moderately in reducing the mortality rate associated with COVID-19 co-infections.

To a more severe extent, the association of low drug response and drug resistance observed among COVID-19 co-infections significantly reduces the recovery rate in these patients and increases associated complications (Lucien et al., 2021; Knight et al., 2021). Moreover, the detrimental effect of several anti-cytokine drugs (to control the cytokine storm) is well reported in literature (Ghasemzadeh et al., 2022; Wilk et al., 2020; Kariyawasam et al., 2022). However, the greatest threats to COVID-19 co-infection are the latent viruses that are activated due to compromised immunity or infection by other viruses. This is well described in cases of CMV and EBV co-infections, where the viruses are reactivated on COVID-19 vaccination. The resulting complication will threaten the safety of millions of individuals by making them vulnerable to more than one (previously uncomplicated) viral infection.

5.5 CONCLUSION

The pathological factors associated with COVID-19 is extensively studied and reported in literature. However, there is lack of documentation related to patients' medical and travel history. The clinical characteristics of co-infections are also rarely documented. This poses critical challenge in analysis of co-infections as compared to single viral infection. Considering similar symptoms and pathological evidence reported in different bacterial as well as viral co-infections, the most effective treatment regimen and ideal management of critically ill patients can be designed only through improved documentation and critical meta-analysis of case studies on COVID-19 co-infections. This strategy may not be very effective during the pandemic, where treatment efforts are desperately needed; yet it may help in controlling the long-term complications arising with COVID-19 co-infections. Through these studies, the similar and variable clinical factors can be determined. Based on this analysis, common features like cytokine storms and liver damage associated with COVID-19 co-infections and the most appropriate therapeutic regimen are now understood and practiced. This has improved the treatment outcome in patients over time. It is also necessary to consider the difference in immunity, exposure and prevalent viral strains in a community. Hence, along with meta-analysis of case studies on COVID-19 co-infections, epidemiological data should also be considered while determining a pattern of endemic infections, their infectivity, opportunistic nature and co-infectivity in COVID-19.

REFERENCES

Aghbash PS, Hemmat N, Nahand JS, Shamekh A, Memar MY, Babaei A, Baghi HB. The role of Th17 cells in viral infections. Int Immunopharmacol. 2021; 91: 107331.

Alhumaid S, Al Mutair A, Al Alawi Z, Alshawi AM, Alomran SA, Almuhanna MS, Almuslim AA, Bu Shafia AH, Alotaibi AM, Ahmed GY, Rabaan AA, Al-Tawfiq JA, Al-Omari A. Coinfections with bacteria, fungi, and respiratory viruses in patients with SARS-CoV-2: A systematic review and meta-analysis. Pathogens. 2021; 10(7): 809.

Angka L, Martel AB, Kilgour M, Jeong A, Sadiq M, de Souza CT, Baker L, Kennedy MA, Kekre N, Auer RC. Natural killer cell IFNγ secretion is profoundly suppressed following colorectal cancer surgery. Ann Surg Oncol. 2018; 25(12): 3747–3754.

Antinori S, Bonazzetti C, Gubertini G. Tocilizumab for cytokine storm syndrome in COVID-19 pneumonia: An increased risk for candidemia? Autoimmun Rev. 2020; 19(7): 102564.

Ardolino M, Azimi CS, Iannello A, Trevino TN, Horan L, Zhang L, Deng W, Ring AM, Fischer S, Garcia KC, Raulet DH. Cytokine therapy reverses NK cell anergy in MHC-deficient tumors. J Clin Invest. 2014; 124(11): 4781–4794.

Babiuk LA, Lawman MJ, Ohmann HB. Viral-bacterial synergistic interaction in respiratory disease. Adv Virus Res. 1988; 35: 219–249.

Bai H, Si L, Jiang A, Belgur C, Zhai Y, Plebani R, Oh CY, Rodas M, Patil A, Nurani A, Gilpin SE, Powers RK, Goyal G, Prantil-Baun R, Ingber DE. Mechanical control of innate immune responses against viral infection revealed in a human lung alveolus chip. Nat Commun. 2022; 13(1): 1928.

Banerjee D, Popoola J, Shah S, Ster IC, Quan V, Phanish M. COVID-19 infection in kidney transplant recipients. Kidney Int. 2020; 97: 1076–1082.

Bhattacharjee S, Banerjee M. Immune thrombocytopenia secondary to COVID-19: A systematic review. SN Compr Clin Med. 2020; 2(11): 2048–2058.

Billah M, Santeusanio A, Delaney V, Cravedi P, Farouk SS. A catabolic state in a kidney transplant recipient with COVID-19. Transpl Int. 2020; 33: 1140–1141.

Blanco JL, Ambrosioni J, Garcia F, Martínez E, Soriano A, Mallolas J, Miro JM. COVID-19 in HIV investigators. COVID-19 in patients with HIV: Clinical case series. Lancet HIV. 2020; 7(5): e314–e316.

Cao X. COVID-19: Immunopathology and its implications for therapy. Nat Rev Immunol. 2020; 20(5): 269–270.

Chaung J, Chan D, Pada S, Tambyah PA. Coinfection with COVID-19 and coronavirus HKU1-The critical need for repeat testing if clinically indicated. J Med Virol. 2020; 92(10): 1785–1786.

Chen N, Zhou M, Dong X, Qu J, Gong F, Han Y, Qiu Y, Wang J, Liu Y, Wei Y, Xia J, Yu T, Zhang X, Zhang L. Epidemiological and clinical characteristics of 99 cases of 2019 novel coronavirus pneumonia in Wuhan, China: A descriptive study. Lancet. 2020a; 395(10223): 507–513.

Chen T, Song J, Liu H, Zheng H, Chen C. Positive Epstein-Barr virus detection in coronavirus disease 2019 (COVID-19) patients. Sci Rep. 2021; 11(1): 10902.

Chen Y, Liu Q, Guo D. Emerging coronaviruses: Genome structure, replication, and pathogenesis. J Med Virol. 2020b; 92(4): 418–423.

Cheng X, Uchida T, Xia Y, Umarova R, Liu CJ, Chen PJ, Gaggar A, Suri V, Mücke MM, Vermehren J, Zeuzem S, Teraoka Y, Osawa M, Aikata H, Tsuji K, Mori N, Hige S, Karino Y, Imamura M, Chayama K, Liang TJ. Diminished hepatic IFN response following HCV clearance triggers HBV reactivation in coinfection. J Clin Invest. 2020; 130(6): 3205–3220.

Chu DKW, Pan Y, Cheng SMS, Hui KPY, Krishnan P, Liu Y, Ng DYM, Wan CKC, Yang P, Wang Q, Peiris M, Poon LLM. Molecular diagnosis of a novel coronavirus (2019-nCoV) causing an outbreak of pneumonia. Clin Chem. 2020; 66: 549–555.

Cuadrado-Payán E, Montagud-Marrahi E, Torres-Elorza M, Bodro M, Blasco M, Poch E, Soriano A, Piñeiro GJ. SARS-CoV-2 and influenza virus co-infection. Lancet. 2020; 395(10236): e84.

Da Cunha T, Mago S, Bath RK. Epstein-Barr virus reactivation causing cholestatic hepatitis. Cureus. 2022; 14(4): e24552.

Dadashi M, Dadashi A, Sameni F, Sayadi S, Goudarzi M, Nasiri MJ, Yaslianifard S, Ghazi M, Arjmand R, Hajikhani B. SARS-CoV-2 and HIV co-infection; clinical features, diagnosis, and treatment strategies: A systematic review and meta-analysis. Gene Rep. 2022; 27: 101624.

Dadashi M, Khaleghnejad S, Abedi Elkhichi P, Goudarzi M, Goudarzi H, Taghavi A, Vaezjalali M, Hajikhani B. COVID-19 and influenza co-infection: A systematic review and meta-analysis. Front Med (Lausanne). 2021; 8: 681469.

Danwang C, Noubiap JJ, Robert A, Yombi JC. Outcomes of patients with HIV and COVID-19 co-infection: A systematic review and meta-analysis. AIDS Res Ther. 2022; 19(1): 3.

D'Ardes D, Boccatonda A, Schiavone C, Santilli F, Guagnano MT, Bucci M, et al. A case of coinfection with SARS-CoV-2 and cytomegalovirus in the era of COVID-19. Eur J Case Rep Intern Med 2020; 7(5): 001652.

Davis B, Rothrock AN, Swetland S, Andris H, Davis P, Rothrock SG. Viral and atypical respiratory co-infections in COVID-19: A systematic review and meta-analysis. J Am Coll Emerg Physicians Open. 2020; 1(4): 533–548.

Ding Q, Lu P, Fan Y, Xia Y, Liu M. The clinical characteristics of pneumonia patients coinfected with 2019 novel coronavirus and influenza virus in Wuhan, China. J Med Virol. 2020; 92: 1549–1555.

Donyavi T, Bokharaei-Salim F, Baghi HB, Khanaliha K, Janat-Makan MA, Karimi B, Nahand JS, Mirzaei H, Khatami A, Garshasbi S, Khoshmirsafa M, Kianib SJ. Acute and post- acute phase of Covid-19: Analyzing expression patterns of miRNA-29a-3p, 146a-3p, 155-5p, and Let-7b-3p in PBMC. Int Immunopharmacol. 2021; 97: 107641.

Dugas M, Grote-Westrick T, Merle U, Fontenay M, Kremer AE, Hanses F, Vollenberg R, Lorentzen E, Tiwari-Heckler S, Duchemin J, Ellouze S, Vetter M, Fürst J, Schuster P, Brix T, Denkinger CM, Müller-Tidow C, Schmidt H, Phil-Robin T, Kühn J. Lack of antibodies against seasonal coronavirus OC43 nucleocapsid protein identifies patients at risk of critical COVID-19. J Clin Virol. 2021; 139: 104847.

Ejaz H, Alsrhani A, Zafar A, Javed H, Junaid K, Abdalla AE, Abosalif KOA, Ahmed Z, Younas S. COVID-19 and comorbidities: Deleterious impact on infected patients. J Infect Public Health. 2020; 13(12): 1833–1839.

Falutz J, Brañas F, Erlandson KM. Frailty: The current challenge for aging people with HIV. Curr Opin HIV AIDS. 2021;16(3):133–140.

Fang Z, Zhang Y, Hang C, Ai J, Li S, Zhang W. Comparisons of viral shedding time of SARS-CoV-2 of different samples in ICU and non- ICU patients. J Infect. 2020; 81(1): 147–178.

Feng Y, Ling Y, Bai T, Xie Y, Huang J, Li J, Xiong W, Yang D, Chen R, Lu F, Lu Y, Liu X, Chen Y, Li X, Li Y, Summah HD, Lin H, Yan J, Zhou M, Lu H, Qu J. COVID-19 with different severities: A multicenter study of clinical features. Am J Respir Crit Care Med. 2020; 201(11): 1380–1388.

Fukuda M, Amano Y, Masumura C, Ogawa M, Inohara H. Development of infectious mononucleosis as an unusual manifestation of COVID-19. Auris Nasus Larynx. 2022; 49(6): 1067–1071.

Geretti AM, Stockdale AJ, Kelly SH, Cevik M, Collins S, Waters L, Villa G, Docherty A, Harrison EM, Turtle L, Openshaw PJM, Baillie JK, Sabin CA, Semple MG. Outcomes of coronavirus disease 2019 (COVID-19) related hospitalization among people with human immunodeficiency virus (HIV) in the ISARIC World Health Organization (WHO) clinical characterization protocol (UK): A prospective observational study. Clin Infect Dis. 2021; 73(7): e2095–e2106.

Ghasemzadeh M, Ghasemzadeh A, Hosseini E. Exhausted NK cells and cytokine storms in COVID-19: Whether NK cell therapy could be a therapeutic choice. Hum Immunol. 2022; 83(1): 86–98.

Guan Z, Chen C, Li Y, Yan D, Zhang X, Jiang D, Yang S, Li L. Impact of coinfection with SARS-CoV-2 and influenza on disease severity: A systematic review and meta-analysis. Front Public Health. 2021; 9: 773130.

Guo L, Wang Y, Kang L, Hu Y, Wang L, Zhong J, Chen H, Ren L, Gu X, Wang G, Wang C, Dong X, Wu C, Han L, Wang Y, Fan G, Zou X, Li H, Xu J, Jin Q, Cao B, Wang J. Cross-reactive antibody against human coronavirus OC43 spike protein correlates with disease severity in COVID-19 patients: A retrospective study. Emerg Microbes Infect. 2021; 10(1): 664–676.

Gushchin VA, Tsyganova EV, Ogarkova DA, Adgamov RR, Shcheblyakov DV, Glukhoedova NV, Zhilenkova AS, Kolotii AG, Zaitsev RD, Logunov DY, Gintsburg AL, Mazus AI. Sputnik V protection from COVID-19 in people living with HIV under antiretroviral therapy. E Clinical Medicine. 2022; 46: 101360.

Han H, Ma Q, Li C, Liu R, Zhao L, Wang W, Zhang P, Liu X, Gao G, Liu F, Jiang Y, Cheng X, Zhu C, Xia Y. Profiling serum cytokines in COVID-19 patients reveals IL-6 and IL-10 are disease severity predictors. Emerg Microbes Infect. 2020; 9(1): 1123–1130.

Hariyanto TI, Rosalind J, Christian K, Kurniawan A. Human immunodeficiency virus and mortality from coronavirus disease 2019: A systematic review and meta-analysis. South Afr J HIV Med. 2021; 22(1): 1220.

Harrison AG, Lin T, Wang P. Mechanisms of SARS-CoV-2 transmission and pathogenesis. Trends Immunol. 2020; 41: 1100–1115.

Hashemi SA, Safamanesh S, Ghasemzadeh-Moghaddam H, Ghafouri M, Azimian A. High prevalence of SARS-CoV-2 and influenza A virus (H1N1) coinfection in dead patients in Northeastern Iran. J Med Virol. 2021; 93(2): 1008–1012.

Hoffmann CJ, Thio CL Clinical implications of HIV and hepatitis B co-infection in Asia and Africa. Lancet Infect Dis. 2007; 7(6): 402–409.

Jackson CB, Farzan M, Chen B. Mechanisms of SARS-CoV-2 entry into cells. Nat Rev Mol Cell Biol. 2022; 23: 3–20.

Ji M, Xia Y, Loo JFC, Li L, Ho HP, He J, Gu D. Automated multiplex nucleic acid tests for rapid detection of SARS-CoV-2, influenza A and B infection with direct reverse-transcription quantitative PCR (dirRT-qPCR) assay in a centrifugal microfluidic platform. RSC Adv. 2020; 10(56): 34088–34098.

Jiang C, Yao X, Zhao Y, Wu J, Huang P, Pan C, Liu S, Pan C. Comparative review of respiratory diseases caused by coronaviruses and influenza A viruses during epidemic season. Microbes Infect. 2020; 22(6–7): 236–244.

Jin CC, Zhu L, Gao C, Zhang S. Correlation between viral RNA shedding and serum antibodies in individuals with coronavirus disease 2019. Clin Microbiol Infect. 2020; 26: 1280–1282.

John ALS, Rathore AP. Early insights into immune responses during COVID-19. J Immunol. 2020; 205: 555–564.

Jothimani D, Venugopal R, Abedin MF, Kaliamoorthy I, Rela M. COVID-19 and the liver. J Hepatol. 2020; 73: 1231–1240.

Ju B, Zhang Q, Ge X, Wang R, Yu J, Shan S, Zhou B, Song S, Tang X, Yu J, Ge J, Lan J, Yuan J, Wang H, Zhao J, Zhang S, Wang Y, Shi X, Liu L, Wang X, Zhang Z, Zhang L. Potent human neutralizing antibodies elicited by SARS-CoV-2 infection. Nature 2020; 584: 115–119.

Kariyawasam JC, Jayarajah U, Abeysuriya V, Riza R, Seneviratne SL. Involvement of the liver in COVID-19: A systematic review. Am J Trop Med Hyg. 2022; 106(4): 1026–1041.

Kim D, Quinn J, Pinsky B, Shah NH, Brown I. Rates of co-infection between SARS-CoV-2 and other respiratory pathogens. JAMA. 2020; 323(20): 2085–2086.

Kimmig LM, Wu D, Gold M. Il-6 inhibition in critically ill COVID-19 patients is associated with increased secondary infections. Front Med. 2020; 7: 583897.

Kiseleva I, Grigorieva E, Larionova N, Al Farroukh M, Rudenko L. COVID-19 in light of seasonal respiratory infections. Biology (Basel). 2020; 9(9): 240.

Knight GM, Glover RE, McQuaid CF, Olaru ID, Gallandat K, Leclerc QJ, Fuller NM, Willcocks SJ, Hasan R, van Kleef E, Chandler CI. Antimicrobial resistance and COVID-19: Intersections and implications. Elife. 2021; 10: e64139.

Krumbein H, Kümmel LS, Fragkou PC, Thölken C, Hünerbein BL, Reiter R, Papathanasiou KA, Renz H, Skevaki C. Respiratory viral co-infections in patients with COVID-19 and associated outcomes: A systematic review and meta-analysis. Rev Med Virol. 2021; 10.1002/rmv.2365.

Kulkarni U, Zemans RL, Smith CA, Wood SC, Deng JC, Goldstein DR. Excessive neutrophil levels in the lung underlie the age-associated increase in influenza mortality. Mucosal Immunol. 2019; 12: 545–554.

Lagunas-Rangel FA, Chávez-Valencia V. High IL-6/IFN-γ ratio could be associated with severe disease in COVID-19 patients. J Med Virol. 2020; 92(10): 1789–1790.

Lai CC, Wang CY, Hsueh PR. Co-infections among patients with COVID-19: The need for combination therapy with non-anti-SARS- CoV-2 agents? J Microbiol Immunol Infect. 2020; 53: 505–512.

Lansbury L, Lim B, Baskaran V, Lim WS. Co-infections in people with COVID-19: A systematic review and meta-analysis. J Infect. 2020; 81(2): 266–275.

Lau SKP, Lung DC, Wong EYM, Aw-Yong KL, Wong ACP, Luk HKH, Li KSM, Fung J, Chan TTY, Tang JYM, Zhu L, Yip CCY, Wong SCY, Lee RA, Tsang OTY, Yuen KY, Woo PCY. Molecular evolution of human coronavirus 229E in Hong Kong and a fatal COVID-19 case involving coinfection with a novel human coronavirus 229e genogroup. mSphere. 2021; 6(1): e00819–e00820.

Lee KW, Yap, SF, Ngeow NF, Lye MS. COVID-19 in people living with HIV: A systematic review and meta-analysis. Int J Environ Res Public Health. 2021; 18(7): 3554.

Léonard-Lorant I, Delabranche X, Séverac F, Helms J, Pauzet C, Collange O, Schneider F, Labani A, Bilbault P, Molière S, Leyendecker P, Roy C, Ohana M. Acute pulmonary embolism in patients with COVID-19 at CT angiography and relationship to d-dimer levels. Radiology. 2020 Sep; 296(3): E189–E191. doi: 10.1148/radiol.2020201561. Epub 2020 Apr 23. PMID: 32324102; PMCID: PMC7233397.

Levi M, Thachil J, Iba T, Levy JH. Coagulation abnormalities and thrombosis in patients with COVID-19. Lancet Haematol. 2020 Jun; 7(6): e438–e440. doi: 10.1016/S2352-3026(20)30145-9. Epub May 11, 2020. PMID: 32407672; PMCID: PMC7213964.

Li M, Wang H, Tian L, Pang Z, Yang Q, Huang T, Fan J, Song L, Tong Y, Fan H. COVID-19 vaccine development: Milestones, lessons and prospects. Signal Transduct Target Ther. 2022; 7(1): 146.

Lin Y, Yuan J, Long Q, Hu J, Deng H, Zhao Z, Chen J, Lu M, Huang A. Patients with SARS-CoV-2 and HBV co-infection are at risk of greater liver injury. Genes Dis. 2021; 8(4): 484–492.

Lippi G, Plebani M, Henry BM. Thrombocytopenia is associated with severe coronavirus disease 2019 (COVID-19) infections: A meta-analysis. Clinica Chimica Acta. 2020; 506: 145–148.

Liu Y, Yang Y, Zhang C, Huang F, Wang F, Yuan J, Wang Z, Li J, Li J, Feng C, Zhang Z, Wang L, Peng L, Chen L, Qin Y, Zhao D, Tan S, Yin L, Xu J, Zhou C, Jiang C, Liu L. Clinical and biochemical indexes from 2019-nCoV infected patients linked to viral loads and lung injury. Sci China Life Sci. 2020; 63(3): 364–374.

Livingston E., Bucher K. Coronavirus disease 2019 (COVID-19) in Italy. JAMA. 2020; 323(14): 1335.

Lu R, Zhao X, Li J, Niu P, Yang B, Wu H, Wang W, Song H, Huang B, Zhu N, Bi Y, Ma X, Zhan F, Wang L, Hu T, Zhou H, Hu Z, Zhou W, Zhao L, Chen J, Meng Y, Wang J, Lin Y, Yuan J, Xie Z, Ma J, Liu WJ, Wang D, Xu W, Holmes EC, Gao GF, Wu G, Chen W, Shi W, Tan W. Genomic characterisation and epidemiology of 2019 novel coronavirus: Implications for virus origins and receptor binding. Lancet. 2020; 395(10224): 565–574.

Lucien MAB, Canarie MF, Kilgore PE, Jean-Denis G, Fénélon N, Pierre M, Cerpa M, Joseph GA, Maki G, Zervos MJ, Dely P, Boncy J, Sati H, Rio AD, Ramon-Pardo P. Antibiotics and antimicrobial resistance in the COVID-19 era: Perspective from resource-limited settings. Int J Infect Dis. 2021; 104: 250–254.

Ma S, Lai X, Chen Z, Tu S, Qin K. Clinical characteristics of critically ill patients co-infected with SARS-CoV-2 and the influenza virus in Wuhan, China. Int J Infect Dis. 2020; 96: 683–687.

Malekifar P, Pakzad R, Shahbahrami R, Zandi M, Jafarpour A, Rezayat SA, Akbarpour S, Shabestari AN, Pakzad I, Hesari E, Farahani A, Soltani S. Viral coinfection among COVID-19 patient groups: An update systematic review and meta-analysis. Biomed Res Int. 2021; 2021: 5313832.

Marjot T, Webb GJ, Barritt AS. COVID-19 and liver disease: Mechanistic and clinical perspectives. Nat Rev Gastroenterol Hepatol. 2021; 18: 348–364.

Marshall NC, Kariyawasam RM, Zelyas N, Kanji JN, Diggle MA. Broad respiratory testing to identify SARS-CoV-2 viral co-circulation and inform diagnostic stewardship in the COVID-19 pandemic. Virol J. 2021; 18(1): 93.

Mazzoni A, Salvati L, Maggi L, Capone M, Vanni A, Spinicci M, Mencarini J, Caporale R, Peruzzi B, Antonelli A, Trotta M, Zammarchi L, Ciani L, Gori L, Lazzeri C, Matucci A, Vultaggio A, Rossi O, Almerigogna F, Parronchi P, Fontanari P, Lavorini F, Peris A, Rossolini GM, Bartoloni A, Romagnani S, Liotta F, Annunziato F, Cosmi L. Impaired immune cell cytotoxicity in severe COVID-19 is IL-6 dependent. J Clin Invest. 2020; 130(9): 4694–4703.

Mehta P, McAuley DF, Brown M, Sanchez E, Tattersall RS, Manson JJ. HLH across speciality collaboration, UK. COVID-19: Consider cytokine storm syndromes and immunosuppression. Lancet. 2020; 395(10229): 1033–1034.

Meng M, Zhang S, Dong X, Sun W, Deng Y, Li W, Li R, Annane D, Wu Z, Chen D. COVID-19 associated EBV reactivation and effects of ganciclovir treatment. Immun Inflamm Dis. 2022; 10(4): e597.

Merani S, Pawelec G, Kuchel GA, McElhaney JE. Impact of aging and cytomegalovirus on immunological response to influenza vaccination and infection. Front Immunol. 2017; 8: 784.

Mirzaei R, Goodarzi P, Asadi M, Soltani A, Aljanabi HAA, Jeda AS, Dashtbin S, Jalalifar S, Mohammadzadeh R, Teimoori A, Tari K, Salari M, Ghiasvand S, Kazemi S, Yousefimashouf R, Keyvani H, Karampoor S. Bacterial co-infections with SARS-CoV-2. IUBMB Life. 2020; 72(10): 2097–2111.

Moon AM, Webb GJ, Aloman C, Armstrong MJ, Cargill T, Dhanasekaran R. High mortality rates for SARS-CoV-2 infection in patients with pre-existing chronic liver disease and cirrhosis: Preliminary results from an international registry. J Hepatol. 2020; 73(3): 705–708.

Motta JC, Gómez CC. Adenovirus and novel coronavirus (SARS- Cov2) coinfection: A case report. IDCases. 2020; 22: e00936.

Musuuza JS, Watson L, Parmasad V, Putman-Buehler N, Christensen L, Safdar N. Prevalence and outcomes of co-infection and superinfection with SARS-CoV-2 and other pathogens: A systematic review and meta-analysis. PLoS One. 2021; 16(5): e0251170.

Nadeem A, Suresh K, Awais H, Waseem S. Epstein-Barr virus coinfection in COVID-19. J Investig Med High Impact Case Rep. 2021; 9: 23247096211040626.

Naendrup JH, Borrega Jorge G, Dennis Alexander E, Alexander S-V, Matthias K, Boris B. Reactivation of EBV and CMV in severe COVID-19—Epiphenomena or trigger of hyperinflammation in need of treatment? A large case series of critically ill patients. J Intensive Care Med. 2021; 0885066621110539.

Nicoli F, Clave E, Wanke K, von Braun A, Bondet V, Alanio C, Douay C, Baque M, Lependu C, Marconi P, Stiasny K, Heinz FX, Muetsch M, Duffy D, Boddaert J, Sauce D, Toubert A, Karrer U, Appay V. Primary immune responses are negatively impacted by persistent herpesvirus infections in older people: Results from an observational study on healthy subjects and a vaccination trial on subjects aged more than 70 years old. E Bio Med. 2022; 76: 103852.

Niitsu T, Shiroyama T, Hirata H, Noda Y, Adachi Y, Enomoto T. Cytomegalovirus infection in critically ill patients with COVID-19. J Infect. 2021; 83: 496–522.

Nomah DK, Reyes-Urueña J, Llibre JM, Ambrosioni J, Ganem FS, Miró JM, Casabona J. HIV and SARS-CoV-2 co-infection: Epidemiological, clinical features, and future implications for clinical care and public health for people living with HIV (PLWH) and Paolucci HIV most-at-risk groups. Curr HIV/AIDS Rep. 2022; 19(1): 17–25.

Paolucci S, Cassaniti I, Novazzi F, Fiorina L, Piralla A, Comolli G, Bruno R, Maserati R, Gulminetti R, Novati S, Mojoli F, Baldanti F. San Matteo Pavia COVID-19 task force. EBV DNA increase in COVID-19 patients with impaired lymphocyte subpopulation count. Int J Infect Dis. 2021; 104: 315–319.

Pascolini S, Granito A, Muratori L, Lenzi M, Muratori P. Coronavirus disease associated immune thrombocytopenia: Causation or correlation? J Microbiol Immunol Infect. 2021; 54(3): 531–533.

Peci A, Tran V, Guthrie JL, Li Y, Nelson P, Schwartz KL, Eshaghi A, Buchan SA, Gubbay JB. Prevalence of co-infections with respiratory viruses in individuals investigated for SARS-CoV-2 in Ontario, Canada. Viruses. 2021; 13(1): 130.

Pinky L, Dobrovolny HM. Coinfections of the respiratory tract: Viral competition for resources. PLoS One. 2016; 11(5): e0155589.

Plüß M, Mese K, Kowallick JT, Schuster A, Tampe D, Tampe B. Case Report: Cytomegalovirus reactivation and pericarditis following ChAdOx1 nCoV-19 vaccination against SARS-CoV-2. Front Immunol. 2022; 12: 784145.

Qin C, Zhou L, Hu Z, et al. Dysregulation of immune response in patients with COVID-19 in Wuhan, China. Clin Infect Dis. 2020; 71: 762–768.

Reche PA. Potential cross-reactive immunity to SARS-CoV-2 from common human pathogens and vaccines. Front Immunol. 2020; 11.

Rockett RJ, Draper J, Gall M, Sim EM, Arnott A, Agius JE, Johnson-Mackinnon J, Fong W, Martinez E, Drew AP, Lee C, Ngo C, Ramsperger M, Ginn AN, Wang Q, Fennell M, Ko D, Hueston L, Kairaitis L, Holmes EC, O'Sullivan MN, Chen SC, Kok J, Dwyer DE, Sintchenko V. Co-infection with SARS-CoV-2 omicron and delta variants revealed by genomic surveillance. Nat Commun. 2022; 13(1): 2745.

Roncati L, Lusenti B, Nasillo V, Manenti A. Fatal SARS-CoV-2 coinfection in course of EBV-associated lymphoproliferative disease. Ann Hematol. 2020; 99(8): 1945–1946.

Russell CD, Fairfield CJ, Drake TM, Turtle L, Seaton RA, Wootton DG, Sigfrid L, Harrison EM, Docherty AB, de Silva TI, Egan C, Pius R, Hardwick HE, Merson L, Girvan M, Dunning J, Nguyen-Van-Tam JS, Openshaw PJM, Baillie JK, Semple MG, Ho A. ISARIC4C investigators. Co-infections, secondary infections, and antimicrobial use in patients hospitalised with COVID-19 during the first pandemic wave from the ISARIC WHO CCP-UK study: A multicentre, prospective cohort study. Lancet Microbe. 2021; 2(8): e354–e365.

Sakamoto H, Ishikane M, Ueda P. Seasonal influenza activity during the SARS-CoV-2 outbreak in Japan. JAMA. 2020; 323(19): 1969–1971.

Samprathi M, Jayashree M. Biomarkers in COVID-19: An up-to-date review. Front Pediatr. 2021; 8: 607647.

Shah S, Shah K, Patel SB, Patel FS, Osman M, Velagapudi P, Turagam MK, Lakkireddy D, Garg J. Elevated D-dimer levels are associated with increased risk of mortality in coronavirus disease 2019: A systematic review and meta-analysis. Cardiol Rev. 2020 Nov/Dec; 28(6): 295–302. doi: 10.1097/CRD.0000000000000330. Epub 2020 Jul 2. PMID: 33017364; PMCID: PMC7437424.

Shen Z, Xiao Y, Kang L, Ma W, Shi L, Zhang L, Zhou Z, Yang J, Zhong J, Yang D, Guo L, Zhang G, Li H, Xu Y, Chen M, Gao Z, Wang J, Ren L, Li M. Genomic diversity of severe acute respiratory syndrome-coronavirus 2 in patients with coronavirus disease 2020. Clin Infect Dis. 2020; 71(15): 713–720.

Simonnet A, Engelmann I, Moreau AS, Garcia B, Six S, El Kalioubie A, Robriquet L, Hober D, Jourdain M. High incidence of Epstein-Barr virus, cytomegalovirus, and human-herpes virus-6 reactivations in critically ill patients with COVID-19. Infect Dis Now. 2021; 51(3): 296–299.

Swets MC, Russell CD, Harrison EM, Docherty AB, Lone N, Girvan M, Hardwick HE, ISARIC4C Investigators, Visser LG, Openshaw PJM, Groeneveld GH, Semple MG. SARS-CoV-2 co-infection with influenza viruses, respiratory syncytial virus, or adenoviruses. Lancet. 2022; 399(10334): 1463–1464.

Tang Y, Liu J, Zhang D, Xu Z, Ji J, Wen C. Cytokine storm in COVID-19: The current evidence and treatment strategies. Front Immunol. 2020; 11: 1708.

Tay MZ, Poh CM, Rénia L, MacAry PA, Ng LFP. The trinity of COVID-19: Immunity, inflammation and intervention. Nat Rev Immunol. 2020; 20(6): 363–374.

Torres Acosta MA, Singer BD. Pathogenesis of COVID-19-induced ARDS: Implications for an ageing population. Eur Respir J. 2020; 56(3): 2002049.

Vizcarra P, Pérez-Elías MJ, Quereda C, et al. Description of COVID-19 in HIV-infected individuals: A single-centre, prospective cohort. The Lancet HIV. 2020; 7: 554.

Vos LM, Bruyndonckx R, Zuithoff NPA, Little P, Oosterheert JJ, Broekhuizen BDL, Lammens C, Loens K, Viveen M, Butler CC, Crook D, Zlateva K, Goossens H, Claas ECJ, Ieven M, Van Loon AM, Verheij TJM, Coenjaerts FEJ, GRACE Consortium. Lower respiratory tract infection in the community: Associations between viral aetiology and illness course. Clin Microbiol Infect. 2021; 27(1): 96–104.

Weber S, Kehl V, Erber J, Wagner KI, Jetzlsperger A-M, Burrell T, Kilian S, Schommers P, Augustin M, Crowell CS, Gerhard M, Winter C, Moosmann A, Spinner CD, Protzer U, Hoffmann D, D'Ippolito E, Busch DH. CMV seropositivity is a potential novel risk factor for severe COVID-19 in non-geriatric patients. PLoS One 2022; 17(5): e0268530.

Wee LE, Ko K, Ho WQ, Kwek G, Tan TT, Wijaya L. Community- acquired viral respiratory infections amongst hospitalized inpatients during a COVID-19 outbreak in Singapore: Co-infection and clinical outcomes. J Clin Virol. 2020; 128: 128.

WHO Coronavirus (COVID-19) Dashboard. https://covid19.who.int/. Accessed June 28, 2022.

Wikby A, Johansson B, Olsson J, Lofgren S, Nilsson BO, Ferguson F. Expansions of peripheral blood CD8 T-lymphocyte subpopulations and an association with cytomegalovirus seropositivity in the elderly: The Swedish NONA immune study. Exp Gerontol. 2002; 37: 445–453.

Wilk AJ, Rustagi A, Zhao NQ, Roque J, Martínez-Colón GJ, McKechnie JL, Ivison GT, Ranganath T, Vergara R, Hollis T, Simpson LJ, Grant P, Subramanian A, Rogers AJ, Blish CA. A single-cell atlas of the peripheral immune response in patients with severe COVID-19. Nat Med. 2020; 26(7): 1070–1076.

Wu C, Chen X, Cai Y, Xia J, Zhou X, Xu S, Huang H, Zhang L, Zhou X, Du C, Zhang Y, Song J, Wang S, Chao Y, Yang Z, Xu J, Zhou X, Chen D, Xiong W, Xu L, Zhou F, Jiang J, Bai C, Zheng J, Song Y. Risk factors associated with acute respiratory distress syndrome and death in patients with coronavirus disease 2019 pneumonia in Wuhan, China. JAMA Intern Med. 2020; 180(7): 934–943.

Wu J, Yu J, Shi X, Li W, Song S, Zhao L, Zhao X, Liu J, Wang D, Liu C, Huang B, Meng Y, Jiang B, Deng Y, Cao H, Li L. Epidemiological and clinical characteristics of 70 cases of coronavirus disease and concomitant hepatitis B virus infection: A multicentre descriptive study. J Viral Hepat. 2021; 28(1): 80–88.

Xia Y, Protzer U. Control of hepatitis B virus by cytokines. Viruses. 2017; 9(1): 1.

Xiong Y, Sun D, Liu Y, Fan Y, Zhao L, Li X, Zhu W. Clinical and high-resolution CT features of the COVID-19 infection: Comparison of the initial and follow-up changes. Invest Radiol. 2020; 55(6): 332–339.

Yang RX, Zheng RD, Fan JG. Etiology and management of liver injury in patients with COVID-19. World J Gastroenterol. 2020; 26: 4753–4762.

Yang S, Wang S, Du M, Liu M, Liu Y, He Y. Patients with COVID-19 and HBV coinfection are at risk of poor prognosis. Infect Dis Ther. 2022; 11(3): 1229–1242.

Yaqinuddin A, Kashir J. Innate immunity in COVID-19 patients mediated by NKG2A receptors, and potential treatment using Monalizumab, Cholroquine, and antiviral agents. Med Hypotheses. 2020; 140: 109777.

Yue H, Zhang M, Xing L, Wang K, Rao X, Liu H, Tian J, Zhou P, Deng Y, Shang J. The epidemiology and clinical characteristics of co-infection of SARS-CoV-2 and influenza viruses in patients during COVID-19 outbreak. J Med Virol. 2020; 92(11): 2870–2873.

Zhang AJ, Lee AC, Chan JF, Liu F, Li C, Chen Y, Chu H, Lau SY, Wang P, Chan CC, Poon VK, Yuan S, To KK, Chen H, Yuen KY. Coinfection by severe acute respiratory syndrome coronavirus 2 and influenza A(H1N1)pdm09 virus enhances the severity of pneumonia in golden syrian hamsters. Clin Infect Dis. 2021; 72(12): e978–e992.

Zhao J, Liao X, Wang H, et al. Early virus clearance and delayed antibody response in a case of COVID-19 with a history of coinfection with HIV-1 and HCV. Clin Infect Dis. 2020a; 71: 2233–2235.

Zhao K, Liu A, Xia Y. Insights into hepatitis B virus DNA integration-55 years after virus discovery. Innovation (Camb). 2020b; 1(2): 100034.

Zheng M, Gao Y, Wang G, Song G, Liu S, Sun D, Xu Y, Tian Z. Functional exhaustion of antiviral lymphocytes in COVID-19 patients. Cell Mol Immunol. 2020; 17(5): 533–535.

Zhu F, Cao Y, Xu S, Zhou M. Co-infection of SARS-CoV-2 and HIV in a patient in Wuhan city, China. J Med Virol. 2020b; 92: 1417–1418.

Zhu X, Ge Y, Wu T, Zhao K, Chen Y, Wu B, Zhu F, Zhu B, Cui L. Co-infection with respiratory pathogens among COVID-2019 cases. Virus Res. 2020a; 285: 198005.

Zubchenko S, Kril I, Nadizhko O, Matsyura O, Chopyak V. Herpesvirus infections and post-COVID-19 manifestations: A pilot observational study. Rheumatol Int. 2022: 1–8.

Part III

COVID-19 and Neurological Complications

6 Neurological Association of COVID-19 and Neurodegenerative Disorders

Shilpa Kumari, Ankit Madeshiya and Sarika Singh

6.1 INTRODUCTION

Novel severe acute respiratory syndrome coronavirus 2 (SARS-CoV-2) is a respiratory syndrome coronavirus, first isolated from three people with pneumonia connected to the cluster of acute respiratory illness cases in Wuhan, China in December 2019. All structural features of the novel SARS-CoV-2 virus particle are similar to the pre-existing coronaviruses in nature; however, this virus induced worldwide pandemic is worrisome and indicates its severity and unknown mechanism of infection. Many nations are devastated due to its contagious nature and acute severe complications particularly related to breathing and multiple organ failure. After its devastating effects worldwide and recognition of symptoms in February 2020, WHO officially named the infection of SARS-CoV2 as a coronavirus disease (COVID-19). The symptoms of COVID-19 are fever, dry cough and fatigue; however, the less common symptoms include sore throat, loss of smell and taste, diarrhoea, headache, conjunctivitis, aches and pains, skin-rashes and also discoloration of fingers or toes. To date, various sequencing data suggest that it contains a single positive strand RNA of 30 kilobase which encodes for ten genes (WHO, 2020). However, recent studies show that SARS-CoV-2 has a high mutation rate (compared to reference strain Wuhan/Hu-1/2019). To date, various strains have emerged with diverse virulence capacity (1,2) and development of broad-spectrum medication is urgently required.

Out of multiple strains, the double mutant delta variant first identified (B.1.617, WHO) in India is supposed to be more contagious and potent to evade the immune recognition, leading to respiratory failure and consequent multiple organ failure and death. Due to a high transmission rate and double mutation, it is posing a threat globally (WHO, 2021) and also raising the concern about the developed

DOI: 10.1201/9781003324911-9

vaccines. Although researchers have worked hard to raise the vaccine, the efficacy-duration studies of these are still in progress. Few studies have already suggested the requirement of booster doses to increase the effectiveness of vaccines. Recently, the new variant omicron (B.1.1.529) that has emerged in South Africa is posing a new threat due to a high mutation rate, compared to other variants and is suggested to be more contagious. Therefore, the generated vaccine might not be effective as the severity and transmissibility of disease is still unclear (WHO) due to insufficient knowledge. New cases are coming up with various clinical complications, and it is becoming difficult to diagnose the COVID-19 except RT-PCR and sequencing analysis. Since the pathological symptoms in infected persons are varied, it may have associations with comorbidities, particularly related to respiratory track and respiratory dysfunction which may worsen the patient's condition.

In this chapter, we discuss the probable correlation of COVID-19 with neurodegenerative disease pathology. The coat of SARS-CoV2 expresses the spike protein with receptor-binding region, which directly binds to extracellular domain of ACE2 and indicates the essential requirement of ACE2 for entry of SARS-CoV2 in cell. (3) In this chapter we discuss the association of COVID-19 and neurodegenerative disorders like Alzheimer's disease (AD), Parkinson's disease (PD) and multiple sclerosis (MS). However, we particularly focus on the PD pathology due to its symptomatic association with COVID-19, like hyposmia and hypogeusia.

6.2 COVID-19 AND NEURODEGENERATIVE DISEASES

In this section we will discuss the broad-spectrum effect of COVID-19 on the neurodegenerative diseases, with a particular focus on the PD, AD and MS depending on the available research articles and reports.

In spite of various reports, it is still unclear whether COVID-19 has direct implications in neurodegenerative disease pathology, and this is due to acute pandemic conditions and research done in a very short time-span. Available findings suggested the presence of virus in brain during infection, though the manifestation of neurological and neuropsychological disease due to this infection is not yet known and needs further exploration. Though few studies have been done, we discuss them here.

Studies have suggested the association of COVID-19 with neurodegenerative diseases and has the potential to infect the brain as well as brainstem. (4) The neurological symptoms such as confusion, headache, hypogeusia/ageusia, hyposmia/anosmia, dizziness, epilepsy, acute cerebrovascular disease is also associated with COVID-19. (5) COVID-19 affects the central nervous system (CNS) through four significant pathways: (a) immediate viral damage of nervous tissue, (b) involvement of immune inflammatory responses which leads to cytokine storm and permeates the blood-brain barrier to cause acute necrotising encephalopathy, (c) unintended host immune response results after an acute infection like Guillain-Barré disorder, though the evidence of such situations and logical consequences are nevertheless unclear and (d) cause systemic sickness. (6) SARS-CoV-2 enters the brain through the olfactory nerve, circulatory system or neuronal pathways. The primary cause is due to the binding of spike proteins of coronavirus to ACE2 receptor, (7) essentially required for virus entry in cell and minimises the lysosomal activity results in accumulation of proteins in neurons hence it may contribute to neurodegeneration diseases. (6) Elevated level of inflammatory responses in COVID-19 may also contribute to the severity of the neurodegenerative disease.

Last year's systemic review suggested that patients with dementia, PD, AD and MS made up a significant portion of hospitalised COVID-19 patients. Such patients suffer with altered mental status and worsen their neurological symptoms. Findings suggested that viral infections with severe neuroinflammation is well associated with the pathogenesis of Alzheimer's disease, Parkinson's disease and MS, suggesting that COVID-19 may have the prospects to incite or accelerate the neurodegeneration. COVID-19 is linked with innate immune response and Cytokines and associated inflammatory mediators consist of interleukin-1β, interleukin-2, interleukin-2 receptor, interleukin-4, interleukin-10, interleukin-18, interferon-γ, C-reactive protein, granulocyte colony-stimulating factor, interferon-γ, CXCL10, monocyte chemoattractant protein 1, macrophage inflammatory protein 1-α and tumour necrosis factor-α are found to be elevated, (8,9) but at the same time, most patients have reduced lymphocyte numbers and exhibit T cell depletion. It is likely that COVID-19 survivors will experience neurodegeneration in the upcoming years because systemic inflammation has been found to encourage cognitive decline and neurodegenerative illness. (10)

6.3 COVID-19 AND AD

COVID-19 and AD share common links with respect to angiotensin-converting enzyme 2 (ACE2) receptors and pro-inflammatory markers, such as interleukin-1 (IL-1), IL-6, cytoskeleton-associated protein 4 (CKAP4), galectin-9 (GAL-9 or Gal-9) and APOE4 allele. (11) The augmented level of ACE2 in AD may further aggravate the COVID-19 related complications in AD patients and worsen the disease pathologies. The data regarding fatality of AD patients is not yet well reported and further studies are required.

A direct connection between AD and ACE2 expression through oxidative stress has been hypothesised. In particular, ageing results in a redox state that is out of balance, producing too many reactive oxygen species (ROS) or the antioxidant system malfunctioning, causing oxidative stress. (12) According to research by Kuo et al., (2020) people with homozygous APOE ε4 (apolipoprotein E4 is more susceptible gene for AD) exhibited a greater prevalence of SARS-CoV-2 infection. In addition, the APOE ε4 ε4 allele was linked to an elevated risk of severe COVID-19, independent of dementia and other comorbidities such type 2 diabetes and cardiovascular disease. (13) As a result, APOE ε4 is a risk factor for both AD and SARS-CoV-2 infection. Since SARS-CoV-2 infection primarily affects the respiratory system and respiratory issues are prevalent in the majority of patients with advanced AD, symptoms worsen when AD patients are infected with SARS-CoV-2. (14) IL-6 is inflammatory marker of COVID-19, notably associated with hippocampal atrophy, (15) which shows cognitive decline. β-amyloid 42 (Aβ42) is a peptide, consists of antimicrobial and antiviral properties but due to dysregulation in inflammatory responses leads to misfolding and accumulation of Aβ42. According to study, extreme inflammation of COVID-19 aggravates the neurodegenerative process by upregulating the NLRP3 (NLR family pyrin domain containing 3), complex of innate immune system, which leads to Aβ42 accumulation and neurofibrillary tangles. (10)

6.4 COVID-19 AND PD

PD pathology and COVID-19 also share a common link, mainly the neuroinflammatory responses. Both innate and adaptive immune responses are involved in COVID-19 infection and PD pathology. PD aetiology has been linked to monocyte chemoattractant protein-1 (MCP-1) and its receptor, CC chemokine receptor-2 (CCR2), which significantly contributes to neuroinflammatory responses. (16) The expression of CCR2 (CC chemokine receptor 2) is particularly high in lung tissue which significantly gets affected in COVID-19 patients. In the bronchoalveolar lavage fluid from the lungs of COVID-19 patients who are undergoing mechanical ventilation, there is increased expression of the CCR2, for monocyte chemoattractant protein (MCP-1), (17) and these monocytes are capable of entering into the brain and probably aggravate the PD pathogenesis or marking the onset of disease. (18) Further evidence is required in this context. The possible link between PD and COVID-19 may also be explained by the overexpression of TLR4 in both patients. Blood cells from COVID-19 and PD patients were reported to have increased levels of TLR4 downstream signalling molecules and inflammasome-related proteins. (19) Particularly the dopaminergic neurons in the nigrostriatal system, which express TLR4, could facilitate the binding of the SARS-CoV-2 spike protein to TLR4 and may serve as a catalyst for neurodegeneration in PD patients. (20) Depression and anxiety are majorly associated with PD symptomology, and this may be due to loss of dopamine, serotonin and other biochemicals, but these facts need further exploration, (21–23) and these symptoms are prevalent in COVID-19 patients due to the infection specific requirement of social isolation and unavailability of caregivers during pandemic. (24)

6.5 COVID-19 AND MS

Multiple sclerosis (MS) is an immune-mediated CNS disorder which requires disease-modifying therapies (DMTs) or immunosuppressives. (25) According to the report of viral isolation from the brains of two MS patients and the electron microscopy detection of coronavirus in an MS brain perivascular immunocyte, coronaviruses have been linked to MS. Previous data also demonstrate that coronaviruses cause demyelination (26) and are capable of inducing T cell-mediated autoimmune reactions in rodents. This lends support to the hypothesis that these viruses play a role in MS pathology. (26) Recently the role of toll-like receptors (TLR) has been suggested in pathogenesis of both MS and COVID-19, primarily through viral particle recognition, activation of the innate immune system and secretion of pro-inflammatory cytokines (27). Another study showed that IL-6 dysregulation, which can impair both innate and acquired immunity, is also associated with the severity of COVID-19 symptoms. Additionally, previous findings have suggested elevated levels of IL-17 in their blood which plays a crucial role in pathogenesis of MS, and in COVID-19 patients also the augmented level of IL-17 was observed, (28) though further in-depth investigation is required in this direction.

6.6 MAJOR ASSOCIATION OF COVID-19 AND PD

Though the association between COVID-19 and neurodegenerative diseases is suggested, the close symptomatic association of COVID-19 with PD was highlighted during the pandemic and cannot be ignored. Therefore, in the following sections the details regarding the alliance between COVID-19 and PD will be discussed.

6.7 RECEPTORS INVOLVED IN COVID-19 AND PD

Loss of dopaminergic neurons and concurrent depleted level of dopamine, along with consequent impaired motor responses in PD patients (29) are well correlated with altered level of dopamine receptors. Previously, these dopamine receptors have been reported as the entry point for the virus into the brain and must be taken into consideration in context to SARS-CoV-2. (30) Succeeding studies have reported that dopamine receptors can enable the entry of viruses like HIV (human immunodeficiency virus), JEV (Japanese encephalitis virus) and dengue, which marks it as serious concern during COVID-19 pandemic. For example, JEV virus exploits dopamine receptor D2 along with phospholipase C to target the dopaminergic neurons, (31) whereas for HIV, the modulation of D1 receptor and CCR5 (chemokine receptor) has been reported. (32) In contrast, it has also been reported that activated dopamine receptors improve the HIV entry into primary human macrophages. (33) Involvement of dopamine receptors in dengue infection has also been suggested as dopamine receptor D4 antagonist inhibits the virus replication through mitogen-activated protein kinase signalling. (34) Effects of influenza A virus on developing dopaminergic neurons has also been suggested. (35) The correlation between SARS-CoV-2 and dopamine receptor has also been suggested; (36) however, much remains to be explored. A study by Cohen et al, 2020 (37) has mentioned a case study on a middle-aged Ashkenazi-Jewish man who probably developed PD after SARS-CoV-2 infection. However, they reported that in spite of anti-SARS-CoV-2 IgG antibodies being present in the serum, the antibodies were not found in the CSF. The patient responded well to the anti-PD medication, pramipexole with quick improvement in clinical signs and symptoms. Since SARS-CoV-2 enters through ACE2 receptor, attention must be paid to the intracellular renin-angiotensin system (RAS) in the context of COVID-19 and PD pathology. High expression level of ACE2 receptors have been reported in the dopamine neurons, and they may act in conjunction with dopamine receptors, though further exploration is still required. Parkinson disorders with depleted dopamine levels displayed reduction in expression level of ACE2 receptors, since ACE2 and DDC (Dopa decarboxylase-enzyme catalysing the biosynthesis of dopamine) are co-expressed and co-regulated in both neuronal and non-neuronal cells, suggestive of their possible pathological association. (38) ACE2 receptors are expressed in dopaminergic neurons, and their expression is significantly decreased during PD conditions. Hence, it is assumed that there may be a credible relation among altered expression level of ACE2 and DDC dysfunction in patients infected with COVID-19 and speculations advocate that PD patients are more vulnerable to coronavirus attack and infection in lungs epithelia. The sizeable interaction between dopamine and RAS have been reported in both the brain and in peripheral tissues. (39) Reports suggested that the coherence of involved receptors must also be explored in both pathologies.

6.8 REDOX HOMEOSTASIS AND INFLAMMATORY RESPONSES IN COVID-19 AND PD

Diverse inflammatory responses are the shared pathological feature in both pathologies; therefore, attention must be paid to the coalition of both peripheral and brain inflammation. Evidence-based observations suggested that the mortality observed in COVID-19 patients is frequently due to uncontrolled immunological responses towards viral infection, rather than organ injury due to viral replication itself. In PD pathogenesis also the neuroinflammation is one of inevitable pathological markers (40,41) with inescapable glial activation. Both innate and adaptive immune responses are involved in COVID-19 infection (42) and PD pathology, (43–45) suggesting the probable exacerbation of immune response that cause the risky immunomodulation in patients and heighten the pathological complications, particularly in COVID-19 (due to acute reactions). Reports suggested that the role of T lymphocytes along with alterations in various cytokines level in PD pathology, (46,47) while in COVID-19 patients both B and T lymphocytes gets activated, the responses are uneven among infected patients and may cause cytokine storm syndrome. (42) In PD the critical role of lymphocytes has been reported as these get infiltrated in the brain as the augmented level of cytokine is observed in the brain and in cerebrospinal fluid of patients, suggesting the critical impact of inflammatory responses in disease onset and progression. (45) Reports suggested the involvement of interferon α, β, ϒ in both COVID-19 infection (48) and PD, (49) exhibiting protective effects. Though COVID-19 is considered as a peripheral infection, studies are coming up indicating that SARS-CoV2 can invade the blood-brain barrier, particularly the spike proteins and need to be explored further. (50,51) However, these findings are worrisome as this may cause severe acute inflammatory reactions and exacerbation of slow progressive PD conditions. These cytokines mediated effects are mostly executed through JAK-STAT, signalling pathways. However, additional studies are required in context of comorbidity of COVID-19 infection and PD.

Reactive oxygen species (ROS) mediated modulation of inflammatory responses has been suggested well in PD pathogenesis. (52) Low level of ROS in T cells is a prerequisite for cell survival while the augmented ROS can affect the T cell receptor signalling, modulate the immune response and can lead to apoptosis or necrosis involving signalling cascades like mitogen-activated protein kinase, phosphoinositide 3-kinase (PI3K)/AKT and JAK/STAT, which could be controlled with antioxidants but needs sequential evaluations for dose regimen. (53,54) The therapeutic effect of antioxidants has been suggested in both COVID-19 infection and PD, but for COVID-19, the mechanistic investigations are limited and need further evaluation. (55–58)

α-synuclein aggregates is one of the pathological hallmarks of PD. It entices the CCR2-positive monocytes and facilitates their infiltration from blood to brain to induce the neuroinflammation. Data showed that COVID-19 patients have strong expression of CCR2 gene in the lung tissue, which promotes inflammation and increased expression of MCP-1 (monocyte chemoattractant protein)- a ligand for CCR2 (59) in the bronchoalveolar lavage fluid during mechanical ventilation. These monocytes are capable of entering into the brain and probably aggravate the PD pathogenesis or marking the onset of disease.

6.9 SENSORY TRAITS IN COVID-19 AND PD

Olfactory dysfunction like hyposmia (reduced smell) or anosmia (absence of smell) and ageusia (diminishing taste) are the non-motor clinical symptoms of PD and are also observed during COVID-19. It has been suggested that anosmia observed in both COVID-19 and PD is due to failure in neurogenesis and no replenishment of the dopaminergic neurons in olfactory sensory neurons in olfactory epithelium. (60) Studies also stated that loss of smell is linked with higher risk of PD onset and observed in about 75% to 95% patients. (61–63) Indeed, hyposmia is one of criterion to differentiate PD and other movement disorders such as multiple system atrophy (MSA), essential tremor (ET), corticobasal degeneration (CBD) and progressive supranuclear palsy (PSP). Findings suggested that hyposmia is generally associated with flaws in cholinergic neuronal functions (limbic archicortex denervation) than nigrostriatal dopaminergic denervation and considerably related with PD specific cognitive impairment. (63) Findings published by Merello et al., 2021 (64) reported three cases of acute PD after COVID-19 infection showed that two patients were suffering from hypertension and asthma, respectively, whereas the third patient was healthy before infection, signifying the varied effects and further exploration is needed. One patient showed the akinetic rigid syndrome while in the other two patients, the clinical asymmetry and tremor with mild respiratory difficulty was found which improved naturally with anti-PD treatments. This case study-based evidence advocated the probable link of SARS-CoV-2 with post or para-infectious PD. However, the data is limited and needs a bigger cohort for statistical significance. PD is associated with irregularities in the olfactory bulb due to its reduced volume or α-synuclein aggregation in the olfactory tract of PD patients. (65) Plentiful dopaminergic cells are located in the glomerular layer of the olfactory bulb that disrupt on PD onset and cause reduced odour perception. (66) Likewise, studies reported that ageusia (taste impairment) during PD pathogenesis was observed in approximately 9% to 54% of patients. (67,68) In contradiction, studies available advocate that such alterations are not PD specific. However, before making a conclusion, the selected sample size and employed measurement methods must be taken into account. The diminished sensitivity for bitter taste has also been observed on the onset of PD, which occurs due to the presence of T2R class of GPCR receptor, which remains present on the tongue to identify the bitter taste. (67) Both nasal and bronchial airways and gut mucosa express T2R form of receptors to identify bacterial invasion and recruit calcium-nitric oxide signalling to rise ciliary beating and avoid bacterial entry. (69) Ageusia is reasonably less frequent than hyposmia in PD patients and further studies are required to comprehend the role of taste in PD. Nevertheless, in COVID-19 it is an evident symptom. Since hyposmia as well as ageusia are the primary clinical symptoms observed in COVID-19 patients, it specifies a strong link with PD pathology. Due to the alike impaired olfactory and taste responses, either PD or COVID-19 can't be ignored as these together may lead to serious repercussions to patient's health, particularly in the case of COVID-19 due to acute respiratory ailments. Therefore, urgent investigations and regulatory guidelines in this direction are required to conduct the related studies.

6.10 METABOLIC AND PSYCHIATRIC SICKNESS IN COVID-19 AND PD

PD patients frequently complain about breathlessness during exertion (35.8%), sputum production (13%) and cough (17.9%). (70) The mortality rate of PD patients who die from pneumonia is greater during PD in comparison to the general population. Such pulmonary dysfunction related signs and symptoms include cough, dyspnea, aspiration, hypoxia, pneumonia, hypercapnia and acute respiratory failure which are also reported in PD patients (71) The chronic consumption of anti-parkinsonian drugs may also induce dyspnea. L-DOPA. induced dyskinesia may trigger of shortness of breath in PD patients, which may be due to lack of muscle control. (72) Subthalamic nucleus deep brain stimulation can also lead to dyspnea, bronchoconstriction, disturbed upper airway obstruction or restriction of chest wall as side effects. (72,73) Pulmonary dysfunction observed in PD patients may be due to a decrease in chest wall compliance and camptocormia have also been reported in patients of advanced PD conditions. (74) Studies showed that PD patients faced extensive weaknesses in both inspiratory and expiratory muscles due to tremor (75) and jerky diaphragm movements, (76) along with impaired lung volumes. (77) Sleep disorders such as REM, insomnia, excessive daytime sleepiness and sleep behaviour disorders are also well connected with PD pathology. (78) Report by Serebrovskaya et al., (1998) (79) showed that impairment in chemoreception during PD conditions lead to low alveolar ventilation and autonomic dysfunction. Since ventilatory function is a main concern in COVID-19 patients and is well connected to the non-motor symptom of PD, the PD patients may be more vulnerable towards SARS-CoV-2 mediated respiratory attacks, and consequent complications must be explored.

In PD, depression and anxiety are well reported psychological ailments with various neurochemical aspects. (80,22) However, if we talk about the biochemical pathophysiology of depression and anxiety during PD conditions, the diagnosed clinical picture is not clear, and no specific etiology is known. Though, it has been suggested that the above discussed symptoms might be due to exhausted level of dopamine, serotonin and norepinephrine, the findings vary in both pre-clinical and clinical studies, thus making it difficult to deliver the statement in this regard. (23,81) Whereas in COVID-19 patients the depression might be due to requirement and guidelines of clinical treatment, as a result of its contagious nature patients mandatorily need to isolate themselves, which makes them feel isolated, and this becomes a probable reason for anxiety and depression. (24,82,83) Regarding the biochemical basis of anxiety and depression in COVID-19 patients, no specific clinical finding is available, and further studies are required. The observed inflammatory responses in both PD and COVID-19 pathologies might be one of the contributors in the observed psychological and metabolic ailment, (84–86) but mechanistic studies are still required in this context.

Clinical observations have suggested that the mental suffering and peripheral inflammatory markers were significantly higher in the infected individuals, which were inversely correlated with blood parameters, including C-reactive protein (CRP), white blood corpuscles (WBCs) count and cytokine levels. (85,86) Advanced therapies such as levodopa-carbidopa intestinal gel (LCIG) implants or deep brain stimulation (DBS) can't be provided through telemedicine to PD patients, who urgently require physical medical attention therefore need to be taken into account for the occurrence of psychological ailments.

6.11 CONCLUDING REMARKS

Aforementioned evidence have clearly suggested the noteworthy association of both COVID-19 and neurodegenerative pathologies, and thus, the comorbidity must be taken into attention. Since patients with neurodegenerative diseases are at higher risk for hospitalisation and death in the setting of COVID-19, additional precautions and protective measures should be put in place to prevent the further infections and comorbidities. Age itself is also one of the critical parameters and must be taken into consideration in both the pathologies of COVID-19 and neurodegenerative disorders. Further the pathological events have some similarities; therefore, the accurate diagnosis is warranted before initiation of treatment which needs further investigation. Besides, the effect of treatment/therapeutics of neurodegenerative disorders on COVID-19 pathology must also be studied in detail to finalise the balanced treatment to patients.

REFERENCES

1. Rfaki A, Touil N, Hemlali M, Alaoui Amine S, Melloul M, El Alaoui MA, Elannaz H, Lahlou AI, Elouennass M, Ennibi K, El Fahime E. Complete genome sequence of a SARS-CoV-2 strain sampled in Morocco in May 2020, obtained using sanger sequencing. Microbiology Resource Announcements. 2021 May 20;10(20):e00387–21.
2. Yang DM, Lin FC, Tsai PH, Chien Y, Wang ML, Yang YP, Chang TJ. Pandemic analysis of infection and death correlated with genomic open reading frame 10 mutation in severe acute respiratory syndrome coronavirus 2 victims. Journal of the Chinese Medical Association. May 1, 2021;84(5):478–84.
3. Liu Z, Xiao X, Wei X, Li J, Yang J, Tan H, Zhu J, Zhang Q, Wu J, Liu L. Composition and divergence of coronavirus spike proteins and host ACE2 receptors predict potential intermediate hosts of SARS-CoV-2. Journal of Medical Virology. 2020 Jun;92(6):595–601.
4. Li YC, Bai WZ, Hashikawa T. The neuroinvasive potential of SARS-CoV2 may play a role in the respiratory failure of COVID-19 patients. Journal of Medical Virology. 2020 Jun;92(6):552–5.
5. Jiménez-Ruiz A, García-Grimshaw M, Ruiz-Sandoval JL. Neurological manifestations of COVID-19. Gaceta Medica de Mexico. May 14, 2020;156(4).
6. Singh AK, Bhushan B, Maurya A, Mishra G, Singh SK, Awasthi R. Novel coronavirus disease 2019 (COVID-19) and neurodegenerative disorders. Dermatologic Therapy. 2020 Jul;33(4):e13591.
7. Ferini-Strambi L, Salsone M. COVID-19 and neurological disorders: Are neurodegenerative or neuroimmunological diseases more vulnerable? Journal of Neurology. 2021 Feb;268(2):409–19.
8. Chen G, Wu D, Guo W, Cao Y, Huang D, Wang H, Wang T, Zhang X, Chen H, Yu H, Zhang X, Zhang M, Wu S, Song J, Chen T, Han M, Li S, Luo X, Zhao J, Ning Q. Clinical and immunological features of severe and moderate coronavirus disease 2019. Journal of Clinical Investigation. May 1, 2020;130(5):2620–9.
9. Mehta P, McAuley DF, Brown M, Sanchez E, Tattersall RS, Manson JJ. COVID-19: Consider cytokine storm syndromes and immunosuppression. The Lancet. 2020 Mar 28;395(10229):1033–4.
10. Heneka MT, Golenbock D, Latz E, Morgan D, Brown R. Immediate and long-term consequences of COVID-19 infections for the development of neurological disease. Alzheimer's Research & Therapy. 2020 Dec;12(1):1–3.
11. Rahman MA, Islam K, Rahman S, Alamin M. Neurobiochemical cross-talk between COVID-19 and Alzheimer's disease. Molecular Neurobiology. 2021 Mar;58(3):1017–23.
12. Ciaccio M, Lo Sasso B, Scazzone C, Gambino CM, Ciaccio AM, Bivona G, Piccoli T, Giglio RV, Agnello L. COVID-19 and Alzheimer's disease. Brain Sciences. 2021 Feb 27;11(3):305.
13. Kuo CL, Pilling LC, Atkins JL, Masoli JA, Delgado J, Kuchel GA, Melzer D. APOE e4 genotype predicts severe COVID-19 in the UK Biobank community cohort. The Journals of Gerontology: Series A. 2020 Nov;75(11):2231–2.
14. Hu C, Chen C, Dong XP. Impact of COVID-19 pandemic on patients with neurodegenerative diseases. Frontiers in Aging Neuroscience. 2021 Apr 8;13:664965.

15. Lindlau A, Widmann CN, Putensen C, Jessen F, Semmler A, Heneka MT. Predictors of hippocampal atrophy in critically ill patients. European Journal of Neurology. 2015;22(2):410–15.
16. Gao L, Tang H, Nie K, Wang L, Zhao J, Gan R, Huang J, Feng S, Zhu R, Duan Z, Zhang Y. MCP-1 and CCR2 gene polymorphisms in Parkinson's disease in a Han Chinese cohort. Neurological Sciences. 2015 Apr;36(4):571–6.
17. Zhou Z, Ren L, Zhang L, Zhong J, Xiao Y, Jia Z, Guo L, Yang J, Wang C, Jiang S, Yang D. Heightened innate immune responses in the respiratory tract of COVID-19 patients. Cell Host & Microbe. 2020 Jun 10;27(6):883–90.
18. Achar A, Ghosh C. COVID-19-Associated Neurological Disorders: The potential route of CNS invasion and blood-brain relevance. Cells. 2020 Oct 27;9(11):2360.
19. Sohn KM, Lee SG, Kim HJ, Cheon S, Jeong H, Lee J, Kim IS, Silwal P, Kim YJ, Paik S, Chung C. COVID-19 patients upregulate toll-like receptor 4-mediated inflammatory signaling that mimic bacterial sepsis. Journal of Korean Medical Science. 2020 Sep 28;35(38).
20. Conte C. Possible link between SARS-CoV-2 infection and Parkinson's disease: The role of toll-like receptor 4. International Journal of Molecular Sciences. 2021 Jul 1;22(13):7135.
21. Marsh L. Depression and Parkinson's disease: Current knowledge. Current Neurology and Neuroscience Reports. 2013 Dec;13(12):1–9.
22. Schrag A, Taddei RN. Depression and anxiety in Parkinson's disease. International Review of Neurobiology. 2017;133:623–55.
23. Burn DJ. Beyond the iron mask: Towards better recognition and treatment of depression associated with Parkinson's disease. Movement Disorders: Official Journal of the Movement Disorder Society. 2002 May;17(3):445–54.
24. Shader RI. COVID-19 and depression. Clinical Therapeutics. 2020 Jun;42(6):962–3.
25. Möhn N, Konen FF, Pul R, Kleinschnitz C, Prüss H, Witte T, Stangel M, Skripuletz T. Experience in multiple sclerosis patients with COVID-19 and disease-modifying therapies: A review of 873 published cases. Journal of Clinical Medicine. 2020 Dec 16;9(12):4067.
26. Murray RS, Brown B, Brain D, Cabirac GF. Detection of coronavirus RNA and antigen in multiple sclerosis brain. Annals of Neurology: Official Journal of the American Neurological Association and the Child Neurology Society. 1992 May;31(5):525–33.
27. Ismail II, Al-Hashel J, Alroughani R, Ahmed SF. A case report of multiple sclerosis after COVID-19 infection: Causality or coincidence? Neuroimmunology Reports. 2021;1:100008, ISSN 2667–257X.
28. Petković F, Castellano B. The role of interleukin-6 in central nervous system demyelination. Neural Regeneration Research. 2016 Dec;11(12):1922–3.
29. Lang AE, Lozano AM. Parkinson's disease. First of two parts. New England Journal of Medicine. 1998 Oct 8;339(15):1044–53.
30. Shaskan EG, Oreland L, Wadell G. Dopamine receptors and monoamine oxidase as virion receptors. Perspectives in Biology and Medicine. 1984;27(2):239–50.
31. Simanjuntak Y, Liang JJ, Lee YL, Lin YL. Japanese encephalitis virus exploits dopamine D2 receptor-phospholipase C to target dopaminergic human neuronal cells. Frontiers in Microbiology. 2017 Apr 11;8:651.
32. Basova L, Najera JA, Bortell N, Wang D, Moya R, Lindsey A, Semenova S, Ellis RJ, Marcondes MC. Dopamine and its receptors play a role in the modulation of CCR5 expression in innate immune cells following exposure to methamphetamine: Implications to HIV infection. PloS One. 2018 Jun 26;13(6):e0199861.

33. Gaskill PJ, Yano HH, Kalpana GV, Javitch JA, Berman JW. Dopamine receptor activation increases HIV entry into primary human macrophages. PloS One. 2014 Sep 30;9(9):e108232.
34. Smith JL, Stein DA, Shum D, Fischer MA, Radu C, Bhinder B, Djaballah H, Nelson JA, Früh K, Hirsch AJ. Inhibition of dengue virus replication by a class of small-molecule compounds that antagonize dopamine receptor d4 and downstream mitogen-activated protein kinase signaling. Journal of Virology. May 15, 2014;88(10):5533–42.
35. Landreau F, Galeano P, Caltana LR, Masciotra L, Chertcoff A, Pontoriero A, Baumeister E, Amoroso M, Brusco HA, Tous MI, Savy VL. Effects of two commonly found strains of influenza A virus on developing dopaminergic neurons, in relation to the pathophysiology of schizophrenia. PloS One. 2012 Dec 10;7(12):e51068.
36. Khalefah MM, Khalifah AM. Determining the relationship between SARS-CoV-2 infection, dopamine, and COVID-19 complications. Journal of Taibah University Medical Sciences. 2020 Dec 1;15(6):550–3.
37. Cohen ME, Eichel R, Steiner-Birmanns B, Janah A, Ioshpa M, Bar-Shalom R, Paul JJ, Gaber H, Skrahina V, Bornstein NM, Yahalom G. A case of probable Parkinson's disease after SARS-CoV-2 infection. The Lancet Neurology. 2020 Oct 1;19(10):804–5.
38. Nataf S. An alteration of the dopamine synthetic pathway is possibly involved in the pathophysiology of COVID-19. Journal of Medical Virology. 2020 Oct;92(10):1743–4.
39. Labandeira-García JL, Garrido-Gil P, Rodriguez-Pallares J, Valenzuela R, Borrajo A, Rodríguez-Perez AI. Brain renin-angiotensin system and dopaminergic cell vulnerability. Frontiers in Neuroanatomy. 2014 Jul 8;8:67.
40. Goswami P, Gupta S, Biswas J, Sharma S, Singh S. Endoplasmic reticulum stress instigates the rotenone induced oxidative apoptotic neuronal death: A study in rat brain. Molecular Neurobiology. 2016 Oct;53(8):5384–400.
41. Jenner P, Olanow CW. Oxidative stress and the pathogenesis of Parkinson's disease. Neurology. 1996 Dec 1;47(6 Suppl 3):161S–70S.
42. Melenotte C, Silvin A, Goubet AG, Lahmar I, Dubuisson A, Zumla A, Raoult D, Merad M, Gachot B, Hénon C, Solary E. Immune responses during COVID-19 infection. Oncoimmunology. 2020 Jan 1;9(1):1807836.
43. Kannarkat GT, Boss JM, Tansey MG. The role of innate and adaptive immunity in Parkinson's disease. Journal of Parkinson's Disease. 2013;3(4):493–514.
44. Schonhoff AM, Williams GP, Wallen ZD, Standaert DG, Harms AS. Innate and adaptive immune responses in Parkinson's disease. Progress in Brain Research. 2020 Jan 1;252:169–216.
45. Hirsch EC, Standaert DG. Ten unsolved questions about neuroinflammation in Parkinson's disease. Movement Disorders. 2021 Jan;36(1):16–24.
46. Baba Y, Kuroiwa A, Uitti RJ, Wszolek ZK, Yamada T. Alterations of T-lymphocyte populations in Parkinson disease. Parkinsonism & Related Disorders. 2005 Dec 1;11(8):493–8.
47. Tansey MG, Goldberg MS. Neuroinflammation in Parkinson's disease: Its role in neuronal death and implications for therapeutic intervention. Neurobiology of Disease. 2010 Mar 1;37(3):510–18.
48. Liu SY, Sanchez DJ, Aliyari R, Lu S, Cheng G. Systematic identification of type I and type II interferon-induced antiviral factors. Proceedings of the National Academy of Sciences. 2012 Mar 13;109(11):4239–44.

49. Ejlerskov P, Hultberg JG, Wang J, Carlsson R, Ambjørn M, Kuss M, Liu Y, Porcu G, Kolkova K, Rundsten CF, Ruscher K. Lack of neuronal IFN-β-IFNAR causes Lewy body-and Parkinson's disease-like dementia. Cell. 2015 Oct 8;163(2):324–39.
50. Rhea EM, Logsdon AF, Hansen KM, Williams LM, Reed MJ, Baumann KK, Holden SJ, Raber J, Banks WA, Erickson MA. The S1 protein of SARS-CoV-2 crosses the blood–brain barrier in mice. Nature Neuroscience. 2021 Mar;24(3):368–78.
51. Zhang L, Zhou L, Bao L, Liu J, Zhu H, Lv Q, Liu R, Chen W, Tong W, Wei Q, Xu Y. SARS-CoV-2 crosses the blood–brain barrier accompanied with basement membrane disruption without tight junctions' alteration. Signal Transduction and Targeted Therapy. 2021 Sep 6;6(1):1–2.
52. Dandekar A, Mendez R, Zhang K. Cross talk between ER stress, oxidative stress, and inflammation in health and disease. Stress Responses. 2015:205–14.
53. Kesarwani P, Murali AK, Al-Khami AA, Mehrotra S. Redox regulation of T-cell function: From molecular mechanisms to significance in human health and disease. Antioxidants & Redox Signaling. 2013 Apr 20;18(12):1497–534.
54. Moro-García MA, Mayo JC, Sainz RM, Alonso-Arias R. Influence of inflammation in the process of T lymphocyte differentiation: Proliferative, metabolic, and oxidative changes. Frontiers in Immunology. 2018 Mar 1;9:339.
55. Filograna R, Beltramini M, Bubacco L, Bisaglia M. Anti-oxidants in Parkinson's disease therapy: A critical point of view. Current Neuropharmacology. 2016 Apr 1;14(3):260–71.
56. Shults CW. Antioxidants as therapy for Parkinson's disease. Antioxidants & Redox Signaling. May 1, 2005;7(5–6):694–700.
57. Jiang T, Sun Q, Chen S. Oxidative stress: A major pathogenesis and potential therapeutic target of antioxidative agents in Parkinson's disease and Alzheimer's disease. Progress in Neurobiology. 2016 Dec 1;147:1–9.
58. Checconi P, De Angelis M, Marcocci ME, Fraternale A, Magnani M, Palamara AT, Nencioni L. Redox-modulating agents in the treatment of viral infections. International Journal of Molecular Sciences. 2020 Jan;21(11):4084.
59. Pairo-Castineira E, Clohisey S, Klaric L, Bretherick AD, Rawlik K, Pasko D, Walker S, Parkinson N, Fourman MH, Russell CD, Furniss J. Genetic mechanisms of critical illness in COVID-19. Nature. 2021 Mar;591(7848):92–8.
60. Rethinavel HS, Ravichandran S, Radhakrishnan RK, Kandasamy M. COVID-19 and Parkinson's disease: Defects in neurogenesis as the potential cause of olfactory system impairments and anosmia. Journal of Chemical Neuroanatomy. 2021 Sep 1;115:101965.
61. Bohnen NI, Müller ML, Kotagal V, Koeppe RA, Kilbourn MA, Albin RL, Frey KA. Olfactory dysfunction, central cholinergic integrity and cognitive impairment in Parkinson's disease. Brain. 2010 Jun 1;133(6):1747–54.
62. Haehner A, Hummel T, Reichmann H. A clinical approach towards smell loss in Parkinson's disease. Journal of Parkinson's Disease. 2014 Jan 1;4(2):189–95.
63. Haehner A, Boesveldt S, Berendse HW, Mackay-Sim A, Fleischmann J, Silburn PA, Johnston AN, Mellick GD, Herting B, Reichmann H, Hummel T. Prevalence of smell loss in Parkinson's disease–a multicenter study. Parkinsonism & Related Disorders. 2009 Aug 1;15(7):490–4.
64. Merello M, Bhatia KP, Obeso JA. SARS-CoV-2 and the risk of Parkinson's disease: Facts and fantasy. The Lancet Neurology. 2021 Feb 1;20(2):94–5.
65. Braak H, del Tredici K. Advances in anatomy embryology and cell biology. Prologue. 2009;201:1–8.

66. Hawkes CH, Del Tredici K, Braak H. Parkinson's disease: A dual-hit hypothesis. Neuropathology and Applied Neurobiology. 2007 Dec;33(6):599–614.
67. Cossu G, Melis M, Sarchioto M, Melis M, Melis M, Morelli M, Tomassini Barbarossa I. 6-n-propylthiouracil taste disruption and TAS2R38 nontasting form in Parkinson's disease. Movement Disorders. 2018 Aug;33(8):1331–9.
68. Doty RL, Hawkes CH. Chemosensory dysfunction in neurodegenerative diseases. Handbook of Clinical Neurology. 2019 Jan 1;164:325–60.
69. Carey RM, Lee RJ. Taste receptors in upper airway innate immunity. Nutrients. 2019 Aug 28;11(9):2017.
70. Lee MA, Prentice WM, Hildreth AJ, Walker RW. Measuring symptom load in Idiopathic Parkinson's disease. Parkinsonism & Related Disorders. 2007 Jul 1;13(5):284–9.
71. Pennington S, Snell K, Lee M, Walker R. The cause of death in idiopathic Parkinson's disease. Parkinsonism & Related Disorders. 2010 Aug 1;16(7):434–7.
72. Rice JE, Antic R, Thompson PD. Disordered respiration as a levodopa-induced dyskinesia in Parkinson's disease. Movement Disorders: Official Journal of the Movement Disorder Society. 2002 May;17(3):524–7.
73. Chalif JI, Sitsapesan HA, Pattinson KT, Herigstad M, Aziz TZ, Green AL. Dyspnea as a side effect of subthalamic nucleus deep brain stimulation for Parkinson's disease. Respiratory Physiology & Neurobiology. 2014 Feb 1;192:128–33.
74. Cardoso SR, Pereira JS. Analysis of breathing function in Parkinson's disease. Arquivos de neuro-psiquiatria. 2002;60:91–5.
75. Brown LK. Respiratory dysfunction in Parkinson's disease. Clinics in Chest Medicine. 1994 Dec;15(4):715–27. PMID: 7867286.
76. Estenne M, Hubert M, De Troyer A. Respiratory-muscle involvement in Parkinson's disease. The New England Journal of Medicine. 1984 Dec 6;311(23):1516–17.
77. Polatli M, Akyol A, Çildağ O, Bayülkem K. Pulmonary function tests in Parkinson's disease. European Journal of Neurology. 2001 Jul 25;8(4):341–5.
78. Shill H, Stacy M. Respiratory complications of Parkinson's disease. In Seminars in respiratory and critical care medicine 2002 (Vol. 23, No. 03, pp. 261–266). Copyright© 2002 by Thieme Medical Publishers, Inc., 333 Seventh Avenue, New York, NY 10001, USA. Tel.:+ 1 (212) 584–4662.
79. Serebrovskaya T, Karaban I, Mankovskaya I, Bernardi L, Passino C, Appenzeller O. Hypoxic ventilatory responses and gas exchange in patients with Parkinson's disease. Respiration. 1998;65(1):28–33.
80. Egan SJ, Laidlaw K, Starkstein S. Cognitive behaviour therapy for depression and anxiety in Parkinson's disease. Journal of Parkinson's Disease. 2015 Jan 1;5(3):443–51.
81. Brigitta B. Pathophysiology of depression and mechanisms of treatment. Dialogues in Clinical Neuroscience. 2002 Mar;4(1):7–20.
82. Pérez-Cano HJ, Moreno-Murguía MB, Morales-López O, Crow-Buchanan O, English JA, Lozano-Alcázar J, Somilleda-Ventura SA. Anxiety, depression, and stress in response to the coronavirus disease-19 pandemic. Cirugia y cirujanos. 2020 Oct;88(5):562–8.
83. Nicolini H. Depression and anxiety in the times of the COVID-19 pandemic. Surgery and Surgeons. 2020 Oct;88(5):542–7.
84. Wang X, Michaelis EK, Foster TC, Foster TC. Selective neuronal vulnerability to oxidative stress in the brain. Frontiers in Aging Neuroscience. 2010;2:1–13.

85. Mazza MG, De Lorenzo R, Conte C, Poletti S, Vai B, Bollettini I, Melloni EM, Furlan R, Ciceri F, Rovere-Querini P, Benedetti F. Anxiety and depression in COVID-19 survivors: Role of inflammatory and clinical predictors. Brain, Behavior, and Immunity. 2020 Oct 1;89:594–600.
86. Guo Q, Zheng Y, Shi J, Wang J, Li G, Li C, Fromson JA, Xu Y, Liu X, Xu H, Zhang T. Immediate psychological distress in quarantined patients with COVID-19 and its association with peripheral inflammation: A mixed-method study. Brain, Behavior, and Immunity. 2020 Aug 1;88:17–27.

Part IV

COVID-19 and Cancer

7 Immuno-Oncological Challenge of COVID-19 in Cancer Care and Prognosis

Atar Singh Kushwah, Shireen Masood, Amreen Shamsad and Monisha Banerjee

7.1 INTRODUCTION

Human beings live in the world of microbes and are closely connected with each other for their survival. Although some are disease-causing, many of them protect us and help in boosting our physiology. Even half of our genetic material originated from viruses. In the case of cancer development, 15% of cancers are caused by chronic viral infection like human papillomavirus (HPV) infection (their persistence caused 80–90% of cervical cancer). Hepatitis B and C virus (HBV & HCV) cause liver fibrosis and cancer (Parkin et al, 1999; Zur Hausen et al, 1991). Inflammation is the primary antiviral response which is responsible for immune cell recruitment and viral clearance. Unfortunately, the persistence of infection leads to chronic inflammation, which is responsible for increased reactive metabolites, oxidative cellular and DNA damage (Banerjee and Vats, 2014). Chronic inflammation is associated with a number of diseases, including diabetes, arthritis, Alzheimer's disease and over a longer period, various types of cancer (Aggarwal et al, 2006). Moreover, cancer development as well as its biological behaviour is shaped by environmental influences such as lifestyle choices or exposures (Aggarwal et al, 2006). In addition to modifying cancer risk, these lifestyle factors cause physiologic changes, increasing the risk for various interrelated comorbidities such as obesity and diabetes; asthma and chronic obstructive pulmonary disease; as well as hypertension, cardiovascular and kidney disease (Moore and Chang, 2010; Ben Neriah and Karin, 2011) These comorbidities can influence cancer treatment (Aggarwal et al, 2006) and have been correlated with adverse outcomes from cancer surgery, impaired ability to deliver effective chemotherapy and radiation toxicity (Chiba et al, 2012).

COVID-19 is an infectious disease caused by severe acute respiratory syndrome coronavirus. The SARS-CoV-2 infection starts with the interaction of spike protein (S) and host cell surface receptor angiotensin-converting enzyme 2 (ACE2) to internalise the virus which is facilitated by transmembrane serine

DOI: 10.1201/9781003324911-11

protease 2 (TMPRSS2) (Hoffmann et al, 2020). Comorbidity includes cancer associated with the severity of the infection which may cause multiple organ failures and deaths (Bakouny et al, 2020). Possibility of developing severe COVID-19 outcomes is higher in individuals from vulnerable populations: patients with metabolic disorders and cancer (Sungnak et al, 2020). However, cancer and COVID-19 have similar pathophysiological events like the cytokine storm, increase oxidative stress and compromised redox (Banerjee and Saxena, 2014; Banerjee and Vats, 2014). Moreover, the clinical relevance of cancer to COVID-19 is based on cytokines, interferons, androgen receptors (AR) and immune signalling molecules (Mehta et al, 2020).

An ecosystem to sustain virus expansion is facilitated through viral genome. Virus structure proteins, proteins involved in replication and the proteins required to disrupt host cell machinery are generated after the entry of virus into the cells (Tan et al, 2020; Mehta et al, 2020). A number of strategies have been developed by viruses to infect their respective hosts which involve molecular mechanisms to hamper pathways meant to prevent their replication, propagation and persistence (Read and Douglas, 2014). Innate immune response through interferon γ mediated pathway helps in the removal of a virus from the host's body. Viruses have also developed proteins to deregulate these immune responses (Grivennikov et al, 2010). Tumour suppressor protein p53 is an important protein playing role in DNA repair, DNA replication, apoptosis, cell proliferation and immune responses; hence controlling several decisions of cells (Yuan et al, 2015; Pampin et al, 2006). This makes p53 an important target for inactivation through mutation or deletion in different cancer (Nguyen et al, 2007). The LT antigen (Large-T) of SV40 (Simian virus 40) was discovered to target p53 in cells through its oncogenic properties. Since then many studies have concluded different molecular devices employed by viruses to control, interrupt, impair or deregulate p53 normal functioning (Yuan et al, 2015). Furthermore, the hallmarks of cancer and COVID-19 have provided a useful conceptual framework for understanding the complex biology of cancer and COVID-19.

7.2 HALLMARKS OF CANCER AND COVID-19

Within the cancer microenvironment, there are various attributes that are unique to cancer cells, such as sustained proliferation, immune-resistance, angiogenesis, metastasis, genomic instability, distorted cellular energetics, immortality, resistance to apoptosis and tumour-promoting chronic inflammation (Figure 7.1). With the onset of COVID-19 infection in cancer patients, some of these features are further promoted to worsen the condition of cancer patients. A recent study concluded a positive correlation between serum cancer biomarkers like human epididymis protein (HE4), D-dimers, Interleukin 6 (IL6), carcinoembryonic antigens (CEA), C-reactive protein (CRP) and carbohydrate antigens (CA242, CA724, CA199) with the pathological progressions of COVID-19 (Wei et al, 2020; Yang et al, 2020). It is also reported that during SARS virus infection,

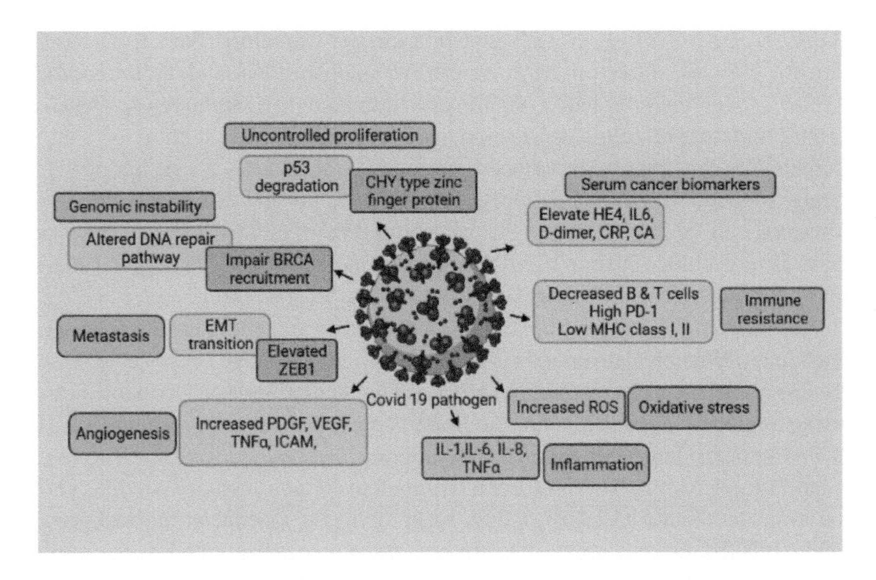

FIGURE 7.1 Emerging hallmarks and characteristics of cancer and COVID-19.

p53 is degraded through non-structural protein of SARS-CoV CHY zinc finger domain and E3 ubiquitin ligase (Ramaiah, 2020). P53 is a tumour suppressor gene that restricts the progression of cell cycle or may induce apoptosis depending on the stage of cell (Shibue et al, 2006). Increased inflammatory responses due to viral infection led to oxidative stress in the cells, which in turn led to activation of p53 and increased mitochondrial outer membrane permeability, causing apoptosis (Shibue et al, 2006). Interestingly, 3a protein of SARS-CoV can bring an inhibition of cell proliferation by limiting the expression of cyclin D3, which may further inhibit Rb phosphorylation (Yuan et al, 2007). It was also found that the spike protein inhibits DNA repair by impairing the recruitment of key DNA repair proteins BRCA1 and 53BP1 to the damage site (Jiang and Mei, 2021). In an experiment performed on SARS-CoV 3CLpro-expressing cells, the amount of cellular reactive oxygen species (ROS) was markedly increased (Lin et al, 2006; Lorente et al, 2021; Salimi et al, 2020). In a different experiment performed on mammalian cells, an increase in ROS was observed (Zhang et al, 2007; Lorente et al, 2021. Elevated ROS leads to oxidative stress in cells which is an important hallmark of cancer cells. COVID-19 causes aberrant STAT3 regulation, which is also observed in cancer cells (Matsuyama et al, 2020). Infection with coronavirus in cancer patients may bring further DNA damage, disruptive signalling and diminished DNA repair ability which may lead to genomic instability in patients with cancer thereby worsening their condition.

The levels of pro-inflammatory cytokines, including IL-6, IL-8, IL-2R, TNF-α are high in cancer patients. The cytokine storm is also an occurrence in patients with severe COVID-19 infections which include inflammatory cytokines like IL-6, IL-1, TNF-α, NF-κB and interferons, worsening the conditions for patients already suffering from cancer (Hantoushzadeh and Norooznezhad, 2020). Cancer patients had a significant reduction in lymphocytes, including CD4+T and CD8+T cells (McDermott and Atkins, 2013). The number of B and T lymphocytes was highly diminished, both CD4+ and CD8+ T cell lymphocytes greatly reduced, the T cell lymphocytes that were present had upregulated programmed cell death 1 Protein (PDCD-1) and downregulated MHC Class I and Class II molecules (Diao et al, 2019). Due to increased PD-1, its ligand PD-L1 binds to it more frequently that downstream signalling of PD-1 receptor in T cells. This inhibits their proliferation and cytotoxic activity, leading to T cell lymphopenia and ultimately cancer cells evading the immune system (Paces et al, 2020). Processes such as inflammation, hypoxia, and oxidative stress are capable of triggering angiogenesis which is an important phenomenon in case of both COVID-19 and cancer (Jahani et al, 2020). A number of pro-angiogenesis factors are found in higher levels among COVID-19 patients which include, vascular endothelial growth factor (VEGF), VEGF receptor 1 (Flt 1), platelet-derived growth factor (PDGF), tumour necrosis factor receptor super family (TNF superfamily), hypoxia-inducible factor 1α (HIF 1α), intracellular adhesion molecule 1 (ICAM 1) and interleukin 6 (IL 6) (Norooznezhad and Mansouri, 2021). In a study conducted upon non-small cell lung cancer (NSCLC) cell line infected with SARS CoV2, the transition of epithelial cells to mesenchymal cells occurs through ACE2 suppression by ZEB1 (a mesenchymal marker) (Stewart et al, 2021).

7.3 GENETICS OF CANCER AND COVID-19

The existence of similar processes and sharing common pathways has led to the upregulation and downregulation of some common genes in cancer and COVID-19. SARS-CoV2 employs ACE2 and TMPRSS2 of the host cells to gain entry into other cells. The expression of both genes was found to be high in colon and stomach cancer patients (Hoang et al, 2021). It might be responsible for high susceptibility to COVID-19 in cancer patients. TMPRSS2 mRNA levels were found higher in colon cancer (Hoang et al, 2021). In prostate cancer, two different studies demonstrated that TMPRSS2 levels are regulated by androgen hormones (Hoang et al, 2021); Lin et al, 1999). In contrast, no change in ACE2 expression was found in lung cancer patients, and a lower expression of TMPRSS2 was found in lung, head and neck cancer patients (Sacconi et al, 2020). The expression of neuropilin 1 (NRP1) and ACE2 were higher in kidney carcinoma cells (Hossain et al, 2021). NRP1 has been recognised as a host cofactor in SARS-CoV2. NOD2 and COX6C expression was found to be (i) downregulated and upregulated in multiple myeloma patients, respectively (ii) downregulated in mild COVID-19 cases (iii) upregulated in severe COVID-19 cases (Wang et al, 2021). Further studies need to be performed to determine the function of these genes in order to draw a concrete conclusion. Interestingly, the C allele mutation in NOD2 gene (rs2111235) was found to lower the risk allele in gastric cancer (Li et al, 2015).

Breast cancer cells express high levels of ATPase H+ Transporting Accessory Protein 1 (ATP6AP1) while, the expression of ATP6AP1 reduces post-SARS CoV2 infection (Wang et al, 2021). EMMPRIN (extracellular matrix metalloproteinase inducer) upregulation is also a common phenomenon in SARS-CoV2 infection and oral cancer (Varadarajan et al, 2020). ERAP2 overexpresses lung carcinomas which have been attributed as a SARS-CoV2 risk factor (Zhang et al, 2021). PTEN, CREB1, CASP3 and SMAD3 are the genes commonly associated with cancer initiation and progression, in silico studies demonstrated their upregulation in SARS-CoV2 as well (Ebrahimi Sadrabadi et al, 2021). The spike protein of SARS-CoV2 upregulates Snail expression, which is also an important protein in epithelial to mesenchymal transition (Lai et al, 2020).

7.4 IMMUNO-ONCOLOGICAL INTERACTION IN CANCER DURING CORONAVIRUS INFECTION

During the early stage of pandemic, the upregulation of many cytokines, including interleukin (IL)-6, IL-1β, tumour necrosis factor α (TNF-α) and interferons was observed in patients with COVID-19 (Huang et al, 2020; McGonagle et al, 2020). An increase in pro-inflammatory cytokines due to immune system diseases leads to a more than normal immune response which leads to a phenomenon known as the cytokine storm (Behrens and Koretzky, 2017). The cytokine storm causes tissue injuries, such as acute lung injury which leads to low saturation levels of oxygen which further causes lung failure. A study demonstrated that almost half of the deaths which were caused by COVID-19 were due to an acute lung injury known as ARDS (acute respiratory distress syndrome) (Liu et al, 2019). More than 50 cytokines have been found to be involved in the cytokine storm and lung injury, the most significant is IL-6, which plays important role in tissue regrowth, immune system functioning metabolic pathways (Kang et al, 2020). IL-6 was found to be involved in both COVID-19 (Wu et al, 2020) and different cancers (Kumari et al, 2016). JAK/STAT signalling pathway is important for cellular mobility, growth and survival, differentiation and activation of immune responses. Therefore, the aberrant JAK/STAT pathway may lead to chronic inflammation resulting in oxidative stress and development of cancer (Johnson et al, 2018).

Within the tumour microenvironment, IL-6 was found to promote the release of a number of pro-inflammatory cytokines which leads to chronic inflammation. Chronic inflammation promotes carcinogenesis. IL-6 has been reported to be a driver for cancer progression, and for cancer diagnosis and prognosis, it has been established as a biomarker (Vargas and Harris, 2016). Along with cytokines, the levels of chemokines like chemotactic proteins (CCL2, CCL3, CCL5, CXCL8, CXCL9, CXCL10) also increases significantly, resulting in an overall deadly inflammatory response (Williams and Chambers, 2014; Channappanavar and Perlman, 2017). The severity of COVID-19 is driven by either excessive proinflammatory cytokines and IL-6 or CD4 lymphopenia-induced lymphocytic dysregulation (Giamarellos-Bourboulis et al, 2020). Increased levels of IL-2, IL-6, IL-7, IL-10, IFNγ and TNF-α are common in patients with severe SARS-CoV2 infection (Huang et al, 2020; Chen et al, 2020; Ruan et al, 2020). Therefore, drug repurposing of anticancer drugs by targeting the IL-6/JAK/STAT pathway/proteins may be saving invaluable time and allow timely delivery for the treatment of COVID-19. Nuclear factor-κB (NF-κB) signalling is the key factor for the production of chemokines and cytokines and is activated by viral genetic materials or proteins (Ma et al, 2020; Hirano and Murakami, 2020).

NF-κB proteins are a class of transcription factors that are widely expressed and mediate the response to cellular stress. They regulate humoral and cellular immune responses by initiating an inflammatory reaction in response to pro-inflammatory factors (Marwarha and Ghribi, 2017). Because of continuous inflammation in tumorigenesis, the NF-κB transcription factor has been closely linked to cancer progression. Since proliferation, differentiation and cell death are all mediated by NF-κB proteins, thus mutations in the gene encoding NF-κB cause abnormal activation of NF-κB transcription factors that have been linked to a variety of malignancies (Carbone and Melisi, 2012). The NF-κB induced inflammatory response and regulated epidermal growth factor (EGF) caused cell death inhibition, metastasis stimulation, proliferation and differentiation of tumour cells (Alberti et al, 2012). Chronic inflammation and oxidative stress disrupt the NF-kB translocation in thymic T cells, making them susceptible to TNF-α mediated apoptosis (Bhattacharyya and Saha, 2015). Pathogens, including viruses like SARS-CoV, can enhance host's cell response to oxidative stress that induce inflammation, resulting in lung injury. NF-kB played a central role in the activation of various inflammatory cytokines (Delgado-Roche and Mesta, 2020).

7.5 OXIDATIVE STRESS AND CHRONIC INFLAMMATION IN CANCER DURING CORONAVIRUS INFECTION

The catalytic domain at the S1/S2 protein present in SARS-CoV-2 makes the virus more virulent. A protease enzyme present on the cell surface known as (TMPRSS2 transmembrane protease serine 2) cleave the complex of S1/S2-ACE2. This further activates spike proteins of SARS-CoV-2, which enables the entry of the virus into the host genome (Ratre et al, 2021; Glowacka et al, 2011). A series of inflammatory processes start after the virus enters into the body; the host's defensive cells are induced by antigen-presenting cells (APCs). Foreign antigen introduced by APCs to CD4+-T-helper (TH1) cells lead to production of interleukin-12 to stimulate TH1. Furthermore, TH1 cells stimulate CD8+-T-killer cells that identify the foreign antigen and destroy contaminated host cells. TH1 cells also stimulate B-cells to produce virus-specific antibodies (Chen et al, 2019; Bennardo et al., 2020). The COVID-19 infection spread through the ACE2 receptors found on nasal mucosa, laryngeal mucosa and lung epithelium of the host (Bennardo et al, 2020).

The C-C Motif Chemokine Ligand 2 (CCL2) is a small chemical molecule that recruits leukocytes such as lymphocytes, monocytes and basophils to the site of infection. It also influences several inflammatory reactions in S-ACE2 signal transduction pathways with diverse inflammatory effects (Hoffmann et al, 2020). A study demonstrated an increased expression of CCL2 (chemotactic protein) in SARS-CoV patients, implying that it may be a potential therapeutic biomarker (Khalil et al, 2022). SARS-CoV covalently attaches to the ACE2 receptor of host cells, the binding causing the dramatic change into ACE2 receptor structure. The binding of the virus and ACE2 receptor mediate downstream signalling involving ERK and AP (Heurich et al, 2014). A study demonstrated that the entry of SARS-CoV is mediated by the interaction of virus surface proteins and lipid bilayer of the host's cell (Simmons et al, 2004).

In clinical biology, the paradox of reactive oxygen species (ROS) is that they are exceedingly harmful and capable of causing a wide range of diseases (Galadari et al, 2017), including respiratory disease, cancer, cardiovascular disease, neurological disease, rheumatoid arthritis and kidney disease. By influencing the host's immunological response, oxidative stress causes immunopathological effects (Channappanavar and Perlman, 2017). COVID-19-related oxidative stress is characterised by decreased adaptive immunity, which leads to severe illness with extensive cellular damage (Alam and Czajkowsky, 2022). Reactive oxygen species (ROS) are molecules with unpaired electrons in their molecular orbitals, which gives them a higher affinity to bind biomolecules (Valko et al, 2007). The respiratory chain in the mitochondrion, which produces a significant number of free radicals during oxidative phosphorylation, is the major source of reactive oxygen species (ROS) formation. Superoxide anion (O_2^-), hydrogen peroxide (H_2O_2), an hydroxyl radicals (OH^-) are the most common (Lupu and Cremer, 2018) and nitric oxide (NO^-) and peroxide nitrite (ONOO–) are the major reactive nitrogen species (RNS) (Poprac et al, 2017). Superoxide dismutase (SOD), catalase (CAT) and glutathione peroxidase (GPx) are the most important antioxidant enzymes which defend the cell from oxidative stress (OS) while ascorbic acid, carotenes glutathione and tocopherols are the major non-enzymatic molecules which maintain the cellular redox (Dinu et al, 2014). Once an inflammatory process is triggered by a specific stimulus, multiple cells are activated and generate various mediators, notably proinflammatory cytokines, which stimulate OS, which is part of the host's first-line defence (Ben-Baruch, 2006). Thus, a prolonged inflammatory reaction develops ROS that is detrimental to cell structures (lipids, proteins and nucleic acids), providing perfect circumstances for cancer development (Federico et al, 2007; Mitran et al, 2018). Chronic OS leads to modification of purine and pyrimidine, purine loss (resulting in abasic sites) and single and double-strand break and cross-links in DNA (Ott et al, 2007). Protein oxidation is primarily caused by OH-attack, which modified the proteins and their accumulation and disrupts cell function by overcoming proteolysis systems. Excessive OS increases protein breakdown and accelerates up to 11 times (Monaghan et al, 2009), for example, carbonyl adducts is the product of protein oxidation (De Marco et al, 2013).

Previous studies reported that malignant cells have a longer lifespan. In most the malignant cells reported hypoxia, excessive growth-promoting molecules and ROS are the primary causes of increased cellular OS (Bhattacharyya and Saha, 2015). ROS appears to operate as the second messenger in intracellular signalling cancer cell persistence (Di Domenico et al, 2012) and makes it more resistant to OS than normal cells (Reuter et al, 2010). Notably, HPV has the flexibility to cope with OS circumstances by the activation of protective molecules such as SOD and CAT (Calaf et al, 2018). These processes appear to be driven by HPV tumorigenesis, particularly E7, which permits invading pathogens to proliferate uncontrollably. Furthermore, E7 regulates the production of Bcl-xL, IL-18, Fas and Bad, leading in cells that are resistant to OS-induced apoptosis (Shim et al, 2008).

De Marco *et al* studied an increase in OS in women samples suffering from cervical dysplastic or neoplastic lesions. OS-induced changes in the structure of proteins related in cell shape and differentiation, such as cytoskeletal keratin 6, actin, cornulin, retinal dehydrogenase and glyceraldehyde 3-phosphate dehydrogenase have been found in dysplastic samples. In contrast, improved regulation of OS was detected in cancer samples, suggesting that OS contributes to the development of a suitable environment for cancer development (De Marco et al, 2012). A significant amount of OS biomarkers are seen in the microenvironment of cancerous cells. Romano et al found a larger amount of 8-hydroxy-2′-deoxyguanosine in dysplastic cells as compared to healthy cells, implying that OS plays an essential role in malignancy (Romano et al, 2000). The investigation on cervical cancer patients by Naidu et al demonstrated a strong link between the MDA (malonyl dialdehyde) level and aggressiveness of lesions, with the highest levels obtained in advanced stages (stage IV) (Naidu et al, 2007). Siegel *et al* proposed an intriguing hypothesis: OS indicators could indicate the immunological host response to HPV infection. They examined the relationships between oxidative stress and infection clearance in 444 HPV-positive women. MDA and anti-5-hydroxymethyl-2′-deoxyuridine auto-antibody (anti-HMdU Ab) levels were determined. They observed elevated levels of two biomarkers that have been linked to quicker clearance of oncogenic HPV strains (Siegel et al, 2012).

Furthermore, it was found that HPV-infected women with high ferritin levels have been less likely to accomplish viral clearance as compared with low ferritin levels. Iron enhances viral multiplication and transcription, while also participating in DNA oxidation, which contributes to ROS production (Siegel et al, 2012). It appears that a lack of specific antioxidants may influence the progress of the infection (Giuliano, 2003) and AP-1 activation leads to the development of E6 and E7 oncoproteins. Manju et al discovered decreased levels of key antioxidant enzymes, GPx, glutathione S-transferase (GST) and SOD, along with vitamin C and E, in individuals with cervical cancer (Sedjo et al, 2002). This could be due to both consumption at site of reaction and sequestering by tumour cells (Manju et al, 2002). Another research on women with cervical intraepithelial neoplasia

(CIN) and cervical squamous cell carcinoma found low SOD and CAT levels but high GPx levels (Looi et al, 2008). Gonçalves *et al* conducted a study of women diagnosed with cervical cancer and precancerous cervical lesions, they found low concentration of vitamin C, a major antioxidant and elevated level of erythrocyte thiobarbituric acid reactive substances (TBARS) and -aminolevulinate dehydratase (-ALA-D). Furthermore, they came to the conclusion that erythrocyte TBARS and vitamin C levels could be employed as OS indicators in the early stages of ovarian cancer. Vitamin C could be beneficial in the therapy of these patients; however, this is still debated (Goncalves et al, 2005).

7.6 IMPACT OF COVID-19 ON TREATMENT OF CANCER

There has been a critical downside in the treatment of cancer patients during the COVID-19 phase, in terms of their susceptibility to COVID-19 and management of the disease during lockdowns. Cancer patients have frequent hospital visits and COVID-19 being an air-borne disease further increased the risk for cancer patients (Chia et al, 2020). Patients with cancer are more susceptible to COVID-19, particularly the ones with haematological and lung cancers who are more likely to develop severe clinical symptoms that would require intensive care unit (ICU) with ventilation and even fatality (Liang et al, 2020; Wu et al, 2020; Zhang et al, 2020). The frequency of lung cancer is probably higher because of the virtue of infection in the same set of cells in the respiratory tract. Patients who are diagnosed with COVID-19 within 14 days of cancer treatment (surgery, chemotherapy, radiotherapy or targeted immune therapy) are likely to develop serious complications (Zhang et al, 2020).

Older age and late-stage cancers (III & IV) were also found to be correlated with higher chances of the COVID-19 infection (Zhang et al, 2020). However, in case of breast cancer, no such correlation was found (Vuagnat et al, 2020). With the progression of tumour development and chronic inflammation, cytokines like IL4, IL5, IL6, IL7 and IL10 start acting like immune suppressing agent(s) (Longbottom et al, 2016). This along with reduced neutrophils and immunosuppressive anti-cancer treatment leads to a weakened immune system, making it difficult for cancer patients to fight COVID-19 infection. In order to manage cancer patients, several measures at different places were applied which include (i) delaying colonoscopy screening and using a non-contact molecular technique in colorectal cancer diagnosis (Liu et al, 2020), (ii) oral anti-cancer drugs instead of intravenous chemotherapy for rectal cancer (Zou et al, 2017), (iii) using an elastometric pump to give chemotherapy at home, (Shereen and Salman, 2019) (iv) postponement of surgeries if possible. Radiation therapy could not be rescheduled; hence cancer patients receiving radiation therapy were treated regularly with great precautionary measures. Some general measures like testing for COVID-19 infection before admitting cancer patients for therapy or surgery, closely monitoring cancer patients receiving anti-tumour therapy for the COVID-19 symptoms and consultation services to be given through electronic communication modes.

Treatment strategies for cancer patients developing COVID-19 infection vary with the type of cancer. For lung cancer patients, testing for COVID-19 is compulsory since the clinical symptoms of both diseases are similar (Banna et al, 2019). COVID-19 testing is required before the immunosuppressive treatment for haematological cancer, gynecological cancers and breast cancer, (Pothuri et al, 2020; Curigliano et al, 2020; Percival et al, 2020). If the result is positive, the possibility of postponement can be considered (Percival et al, 2020). In the case of leukaemia, myeloid suppressors are used for treatment. Oral anti-cancerous drug usage is recommended in case of COVID-19 infection (Paul et al, 2021).

7.7 PROGNOSIS AND PREVENTIVE APPROACHES FOR CANCER AND COVID-19

Cancer patients are at a higher risk of developing COVID-19 and are more likely to have severe clinical outcomes like higher mortality rates (Figure 7.2). This

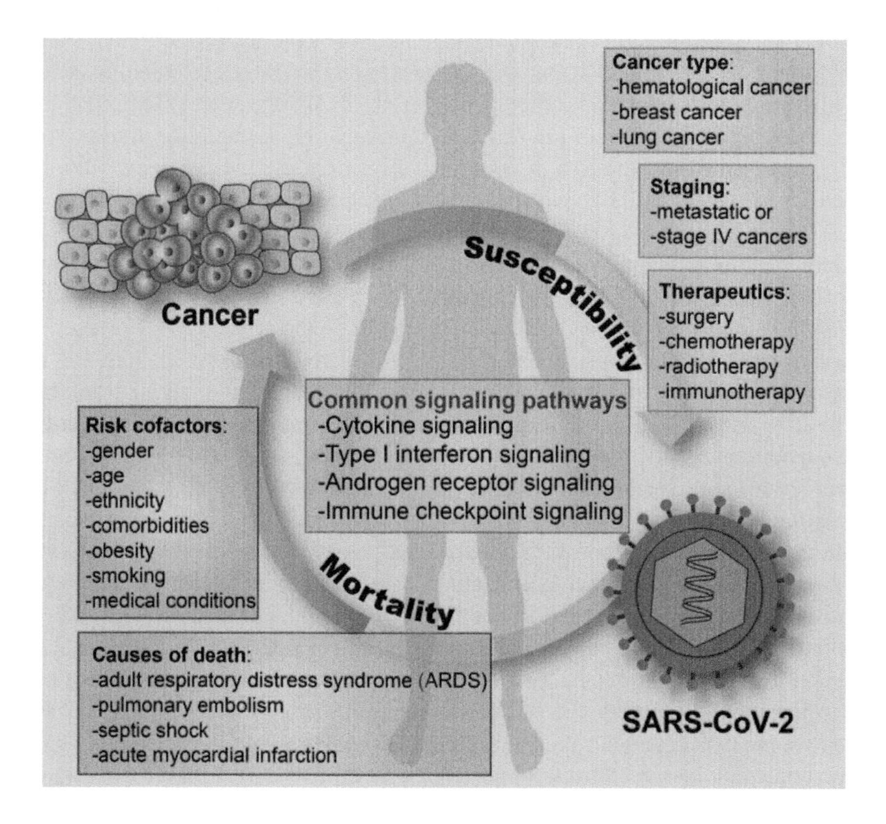

FIGURE 7.2 Crosstalk between COVID-19 and cancer (Zong et al, 2021).

brings up the need for specialised caretaking of the cancer patients. Common strategies have undertaken to care for cancer during the COVID-19 pandemic, 1) delaying chemotherapy or elective surgeries in stable patients; 2) increased personal precautions for all cancer patients; and 3) intensive treatment to cancer patients who were infected with SARS-CoV2. The treatment of cancer patients is recommended to be continued despite the COVID-19 pandemic. Delaying of surgeries can be done in some cases, and vaccination is promoted and prioritised for all cancer patients at all stages.

Different types of cancers have different effects on COVID-19 severity. Patients with blood cancers and breast cancers were more vulnerable to SARS-CoV2 infection and positively correlated with high hospitalisation and mortality rates (Rugge et al, 2020). These patients showed longer course of SARS-CoV2 infection and poor response to the treatment than those without the COVID-19 infection (Passamonti et al, 2020). Apart from cancer types, cancer staging has also been reported to influence COVID-19 infection. The stage of cancer whether metastatic stage or initial stage has also been reported to influence COVID-19 outcomes (Robilotti et al, 2020). Many therapeutic strategies practiced in treatment of cancers like radiotherapy, immunotherapy, chemotherapy and surgeries have been linked to poor diagnosis and mortality in COVID-19 outcomes among cancer patients (Derosa et al, 2020).

A population-based study performed in Wuhan hospitals revealed that out of the 28 patients with cancer and COVID-19, 15 patients had severe complications, and 8 resulted in mortality (Zhang et al, 2020). Such high rates in mortality were also reported in patients with haematological cancer with concurrent COVID-19 in China as well as in Italy (Guan et al, 2020; He et al, 2020). Moreover, another study conducted in Italy demonstrated that 9.5% of men who were diagnosed with COVID-19 had cancer. Intriguingly, men who had cancer were more likely to die from diagnosis of COVID-19 (Montopoli et al, 2020). A large study showed that patients with cancer over 65 years old, those with certain comorbidities, and those who received cancer-directed therapy harbour the greatest risk of death, further supporting the hypothesis that patients with cancer are at increased risk of mortality (Lunski et al, 2021). Conversely, a large group of SARS-CoV2 infected patients who developed cancer in New York City showed no significant differences in mortality compared to non-cancer patients (Miyashita et al, 2020). However, gender, age, ethnicity, obesity, smoking and various medical conditions have been reported to have a tremendous impact on the mortality rates of patients with COVID-19 (Passamonti et al, 2020).

Due to weak immunity and older age, cancer patients are more prone to develop severe clinical symptoms when infected with SARS-CoV2. A lot of preventive measures ranging from hospital and outpatient department (OPD) rules to chemotherapy and radiotherapy management for cancer patients were adopted to safeguard them from SARS-CoV2 infection (van de Haar et al, 2020; Fortin et al, 2022; Fong et al, 2021).

Preventive measures adopted in hospitals and OPDs include the following:

- Temperature testing at the hospital and OPD entrance to ensure restricted entry
- Instructing cancer patients to avoid hospital visits if the SARS-CoV2 symptoms appear (unless there is an emergency)
- Switching to electronic communication for management of cancer-related symptoms and medications
- Recording contact and travel history of cancer patients who had to compulsorily consult with doctors and visit the hospital
- Scheduling appointments online to avoid crowding
- Restricting pre-clinical research work completely and patients involved in clinical trials

Preventive measures adopted in treatment of cancer patients,

- Medications delivery at home
- Switching to oral medication and subcutaneous treatment
- The blood test to be performed at local pathologies rather than visiting hospitals
- The reduction in intravenous treatment for non-aggressive cancers
- The replacement of surgery with radiation therapy, if possible
- Postponement of surgeries, if possible
- Follow-up checkups to be outsourced to local clinics

7.8 CONCLUSION

The death toll resulting from widespread COVID-19 pandemic is high among some portions of the population particularly patients with cancer. Most importantly, the type of cancer, stage of cancer, and therapeutics affect both the incidence and prevalence of COVID-19 infection. Currently there are conflicting results regarding the correlation between mortality rates of patients with cancer and COVID-19. Further studies are required to determine whether cancer per se is an independent risk factor for developing COVID-19. Treating patients with both cancer and COVID-19 is difficult during this period, raising a desperate quest for the treatment that kills two birds with one stone. With joint efforts, we have accumulated an unprecedented perspective on the clinical relevance and molecular interactions governing the incidence and severity of both diseases.

ACKNOWLEDGEMENTS

The authors extend their gratitude to the Indian Council of Medical Research (ICMR), New Delhi, University Grant Commission (UGC), New Delhi, India and Centre of Excellence, higher education, Government of Uttar Pradesh, Lucknow, India for financial support. ASK acknowledges ICMR for senior research fellowship. SM and AS acknowledges CSIR and UGC-MANF for their respective junior research fellowships.

REFERENCES

Aggarwal, B. B., Shishodia, S., Sandur, S. K., *et al* (2006). Inflammation and cancer, how hot is the link? *Biochemical Pharmacology*, 72(11), 1605–1621.

Alam, M. S. and Czajkowsky, D. M. (2022). SARS-CoV-2 infection and oxidative stress, Pathophysiological insight into thrombosis and therapeutic opportunities. *Cytokine and Growth Factor Reviews*, 63, 44–57.

Alberti, C., Pinciroli, P., Valeri, B., *et al* (2012). Ligand-dependent EGFR activation induces the co-expression of IL-6 and PAI-1 via the NFkB pathway in advanced-stage epithelial ovarian cancer. *Oncogene*, 31(37), 4139–4149.

Arciello, M., Gori, M., Balsano, C., *et al* (2013). Mitochondrial dysfunctions and altered metals homeostasis, new weapons to counteract HCV-related oxidative stress. *Oxidative Medicine and Cellular Longevity*, 971024.

Bakouny, Z., Hawley, J. E., Choueiri, T. K., *et al* (2020). COVID-19 and cancer, current challenges and perspectives. *Cancer Cell*, 38(5), 629–646.

Banerjee, M. and Saxena, M. (2014). Genetic polymorphisms of cytokine genes in type 2 diabetes mellitus. *World Journal of Diabetes*, 5(4), 493–504.

Banerjee, M. and Vats, P. (2014). Reactive metabolites and antioxidant gene polymorphisms in type 2 diabetes mellitus. *Redox Biology*, 2, 170–177.

Banna, G., Curioni-Fontecedro, A., Friedlaender, A., *et al* (2019). How we treat patients with lung cancer during the SARS-CoV-2 pandemic, primum non nocere. *ESMO Open*, 5(2), e000765.

Behrens, E. M. And Koretzky, G. A. (2017). Cytokine storm syndrome, looking toward the precision medicine era. *Arthritis and Rheumatology*, 69(6), 1135–1143.

Ben-Baruch, A. (2006). Inflammation-associated immune suppression in cancer, the roles played by cytokines, chemokines and additional mediators. *Seminars in Cancer Biology, Academic Press*, 16(1), 38–52.

Ben-Neriah, Y. and Karin, M. (2011). Inflammation meets cancer, with NF-κB as the matchmaker. *Nature Immunology*, 12(8), 715–723.

Bennardo, F., Buffone, C. and Giudice, A. (2020). New therapeutic opportunities for COVID-19 patients with Tocilizumab, Possible correlation of interleukin-6 receptor inhibitors with osteonecrosis of the jaws. *Oral Oncology*, 106, 104659.

Bhattacharyya, S. and Saha, J. (2015). Tumour, oxidative stress and host T cell response, cementing the dominance. *Scandinavian Journal of Immunology*, 82(6), 477–488.

Calaf, G. M., Urzua, U., Termini, L., *et al* (2018). Oxidative stress in female cancers. *Oncotarget*, 9(34), 23824–23842.

Carbone, C. and Melisi, D. (2012). NF-κB as a target for pancreatic cancer therapy. *Expert Opinion on Therapeutic Targets*, 16(sup2), S1–0.

Channappanavar, R. and Perlman, S. (2017). Pathogenic human coronavirus infections, causes and consequences of cytokine storm and immunopathology. *Seminars in Immunopathology. Springer Berlin Heidelberg*, 39(5), 529–539.

Chen, N., Zhou, M., Dong, X., *et al* (2020). Epidemiological and clinical characteristics of 99 cases of 2019 novel coronavirus pneumonia in Wuhan, China, a descriptive study. *The Lancet*, 395(10223), 507–513.

Chia, P. Y., Coleman, K. K., Tan, Y. K., *et al* (2020). Detection of air and surface contamination by SARS-CoV-2 in hospital rooms of infected patients. *Nature Communications*, 11(1), 1–7.

Chiba, T., Marusawa, H. and Ushijima, T. (2012). Inflammation-associated cancer development in digestive organs, mechanisms and roles for genetic and epigenetic modulation. *Gastroenterology*, 143(3), 550–563.

Curigliano, G., Cardoso, M. J., Poortmans, P., *et al* (2020). Recommendations for triage, prioritization and treatment of breast cancer patients during the COVID-19 pandemic. *The Breast*, 52, 8–16.

De Marco, F. (2013). Oxidative stress and HPV carcinogenesis. *Viruses*, 5(2), 708–731.

De Marco, F., Bucaj, E., Foppoli, C., *et al* (2012). Oxidative stress in HPV-driven viral carcinogenesis, redox proteomics analysis of HPV-16 dysplastic and neoplastic tissues. *PloS one*, 7(3), e34366.

Delgado-Roche, L. and Mesta, F. (2020). Oxidative stress as key player in severe acute respiratory syndrome coronavirus (SARS-CoV) infection. *Archives of Medical Research*, 51(5), 384–387.

Derosa, L., Melenotte, C., Griscelli, F., *et al* (2020). The immuno-oncological challenge of COVID-19. *Nature Cancer*, 1(10), 946–964.

Di Domenico, F., Foppoli, C., Coccia, R., *et al* (2012). Antioxidants in cervical cancer, chemopreventive and chemotherapeutic effects of polyphenols. *Biochimica et Biophysica Acta (BBA)-Molecular Basis of Disease*, 1822(5), 737–747.

Diao, B., Wang, C., Tan, Y., *et al* (2020). Reduction and functional exhaustion of T cells in patients with coronavirus disease 2019 (COVID-19). *Frontiers in Immunology*, 11, 827.

Dinu, LU., Ene, CD., Nicolae, IL., *et al* (2014). The serum levels of 8-hidroxydeoxyguanosine under the chemicals influence. *Revista de Chimie*, 65, 1319–1326.

Ebrahimi Sadrabadi, A., Bereimipour, A., Jalili, A., *et al* (2021). The risk of pancreatic adenocarcinoma following SARS-CoV family infection. *Scientific Reports*, 11(1), 1–13.

Federico, A., Morgillo, F., Tuccillo, C., *et al* (2007). Chronic inflammation and oxidative stress in human carcinogenesis. *International Journal of Cancer*, 121(11), 2381–2386.

Fong, D., San Nicolò, K. O., Alber, M., *et al* (2021). Evaluating the longitudinal effectiveness of preventive measures against COVID-19 and seroprevalence of IgG antibodies to SARS-CoV-2 in cancer outpatients and healthcare workers. *Wiener Klinische Wochenschrift*, 133(7), 359–363.

Fortin, J., Rivest-Beauregard, M., Defer, C., *et al* (2022). The impact of canadian medical delays and preventive measures on breast cancer experience, a silent battle masked by the COVID-19 pandemic. *Canadian Journal of Nursing Research*, 08445621221097520.

Galadari, S., Rahman, A., Pallichankandy, S., *et al* (2017). Reactive oxygen species and cancer paradox, to promote or to suppress? *Free Radical Biology and Medicine*, 104, 144–164.

Georgescu, S. R., Ene, C. D., Tampa, M., *et al* (2016). Oxidative stress-related markers and alopecia areata through latex turbidimetric immunoassay method. *Materiale Plastice*, 53, 522–526.

Giamarellos-Bourboulis, E. J., Netea, M. G., Rovina, N., *et al* (2020). Complex immune dysregulation in COVID-19 patients with severe respiratory failure. *Cell Host & Microbe*, 27(6), 992–1000.

Giuliano, A. (2003). Cervical carcinogenesis, the role of co-factors and generation of reactive oxygen species. *Salud Publica de Mexico*, 45(S3), 354–360.

Glowacka, I., Bertram, S., Müller, M.A., *et al* (2011). Evidence that TMPRSS2 activates the severe acute respiratory syndrome coronavirus spike protein for membrane fusion and reduces viral control by the humoral immune response. *Journal of Virology*, 85, 4122–4134.

Gonçalves, T. L., Erthal, F., Corte, C. L., *et al* (2005). Involvement of oxidative stress in the pre-malignant and malignant states of cervical cancer in women. *Clinical Biochemistry*, 38(12), 1071–1075.

Grivennikov, S. I., Greten, F. R. and Karin, M. (2010). Immunity, inflammation, and cancer. *Cell*, 140(6), 883–899.

Guan, W. J., Ni, Z. Y., Hu, Y., *et al* (2020). Clinical characteristics of coronavirus disease 2019 in China. *New England Journal of Medicine*, 382(18), 1708–1720.

Hantoushzadeh, S. and Norooznezhad, A. H. (2020). Possible cause of inflammatory storm and septic shock in patients diagnosed with (COVID-19). *Archives of Medical Research*, 51(4), 347–348.

He, W., Chen, L., Chen, L., *et al* (2020). COVID-19 in persons with haematological cancers. *Leukemia*, 34(6), 1637–1645.

Heurich, A., Hofmann-Winkler, H., Gierer, S., *et al* (2014). TMPRSS2 and ADAM17 cleave ACE2 differentially and only proteolysis by TMPRSS2 augments entry driven by the severe acute respiratory syndrome coronavirus spike protein. *Journal of Virology*, 88(2), 1293–1307.

Hirano, T. and Murakami, M. (2020). COVID-19, a new virus, but a familiar receptor and cytokine release syndrome. *Immunity*, 52(5), 731–733.

Hoang, T., Nguyen, T. Q. and Tran, T. T. A. (2021). Genetic susceptibility of ACE2 and TMPRSS2 in six common cancers and possible impacts on COVID-19. *Cancer Research and Treatment, Official Journal of Korean Cancer Association*, 53(3), 650–656.

Hoffmann, M., Kleine-Weber, H., Schroeder, S., *et al* (2020). SARS-CoV-2 cell entry depends on ACE2 and TMPRSS2 and is blocked by a clinically proven protease inhibitor. *Cell*, 181(2), 271–280.

Hossain, M. G., Akter, S. and Uddin, M. J. (2021). Emerging role of neuropilin-1 and angiotensin-converting enzyme-2 in renal carcinoma-associated COVID-19 pathogenesis. *Infectious Disease Reports*, 13(4), 902–909.

Huang, C., Wang, Y., Li, X., *et al* (2020). Clinical features of patients infected with 2019 novel coronavirus in Wuhan, China. *The Lancet*, 395(10223), 497–506.

Hulbert, A. J., Faulks, S. C., Buffenstein, R., *et al* (2006). Oxidation-resistant membrane phospholipids can explain longevity differences among the longest-living rodents and similarly-sized mice. *The Journals of Gerontology Series A, Biological Sciences and Medical Sciences*, 61(10), 1009–1018.

Jahani, M., Rezazadeh, D., Mohammadi, P., *et al* (2020). Regenerative medicine and angiogenesis: Challenges and opportunities. *Advanced Pharmaceutical Bulletin*, 10(4), 490.

Jiang, H. and Mei, Y. F. (2021). SARS–CoV–2 spike impairs DNA damage repair and inhibits V (D) J recombination in vitro. *Viruses*, 13(10), 2056.

Johnson, D. E., O'Keefe, R. A. and Grandis, J. R. (2018). Targeting the IL-6/JAK/STAT3 signalling axis in cancer. *Nature Reviews Clinical Oncology*, 15(4), 234–248.

Kang, S., Tanaka, T., Narazaki, M., *et al* (2020). Targeting interleukin-6 signaling in clinic. *Immunity*, 50(4), 1007–1023.

Kohen, R. and Nyska, A. (2002). Invited review, Oxidation of biological systems, oxidative stress phenomena, antioxidants, redox reactions, and methods for their quantification. *Toxicologic Pathology*, 30(6), 620–650.

Khalil, B. A., Shakartalla, S. B., Goel, S., *et al* (2022). Immune profiling of COVID-19 in correlation with SARS and MERS. *Viruses*, 14(1), 164.

Kumari, N., Dwarakanath, B. S., Das, A., *et al* (2016). Role of interleukin-6 in cancer progression and therapeutic resistance. *Tumor Biology*, 37(9), 11553–11572.

Lai, C. C., Shih, T. P., Ko, W. C., *et al* (2020). Severe acute respiratory syndrome coronavirus 2 (SARS-CoV-2) and coronavirus disease-2019 (COVID-19), The epidemic and the challenges. *International Journal of Antimicrobial Agents*, 55(3), 105924.

Li, Z. X., Wang, Y. M., Tang, F. B., *et al* (2015). NOD1 and NOD2 genetic variants in association with risk of gastric cancer and its precursors in a Chinese population. *PloS One*, 10(5), e0124949.

Liang, W., Guan, W., Chen, R., *et al* (2020). Cancer patients in SARS-CoV-2 infection, a nationwide analysis in China. *The Lancet Oncology*, 21(3), 335–337.

Lin, B., Ferguson, C., White, J. T., *et al* (1999). Prostate-localized and androgen-regulated expression of the membrane-bound serine protease TMPRSS2. *Cancer Research*, 59(17), 4180–4184.

Lin, C. W., Lin, K. H., Hsieh, T. H., *et al* (2006). Severe acute respiratory syndrome coronavirus 3C-like protease-induced apoptosis. *FEMS Immunology and Medical Microbiology*, 46(3), 375–380.

Liou, G. Y. and Storz, P. (2010). Reactive oxygen species in cancer. *Free Radical Research*, 44(5), 479–496.

Liu, C., Zhao, Y., Okwan-Duodu, D., *et al* (2020). COVID-19 in cancer patients, risk, clinical features, and management. *Cancer Biology & Medicine*, 17(3), 519.

Liu, Y., Sun, W., Li, J., *et al* (2019). Clinical features and progression of acute respiratory distress syndrome in coronavirus disease. *MedRxiv*. https://doi.org/10.1101/2020.02.17.20024166.

Looi, M. L., Dali, A. Z., Ali, S. A., *et al* (2008). Oxidative damage and antioxidant status in patients with cervical intraepithelial neoplasia and carcinoma of the cervix. *European Journal of Cancer Prevention*, 17(6), 555–560.

Longbottom, E. R., Torrance, H. D., Owen, H. C., *et al* (2016). Features of postoperative immune suppression are reversible with interferon gamma and independent of interleukin-6 pathways. *Annals of Surgery*, 264(2), 370–377.

Lorente, L., Martín, M. M., González-Rivero, A. F., *et al* (2021). DNA and RNA oxidative damage and mortality of patients with COVID-19. *The American Journal of Medical Sciences*, 361(5), 585–590.

Lunski, M. J., Burton, J., Tawagi, K., *et al* (2021). Multivariate mortality analyses in COVID-19, comparing patients with cancer and patients without cancer in Louisiana. *Cancer*, 127(2), 266–274.

Lupu, A. R. and Cremer, L. (2018). Hydroxyl radical scavenger activity of natural SOD. *Romanian Archives of Microbiology and Immunology*, 77(1), 73–87.

Ma, Q., Pan, W., Li, R., *et al* (2020). Liu Shen capsule shows antiviral and anti-inflammatory abilities against novel coronavirus SARS-CoV-2 via suppression of NF-κB signaling pathway. *Pharmacological Research*, 158, 104850.

Manju, V., Sailaja, J. K., Nalini, N., *et al* (2002). Circulating lipid peroxidation and antioxidant status in cervical cancer patients, a case-control study. *Clinical Biochemistry*, 35(8), 621–625.

Marwarha, G. and Ghribi, O. (2017). Nuclear factor kappa-light-chain-enhancer of activated B cells (NF-KB)–a friend, a foe, or a bystander-in the neurodegenerative cascade and pathogenesis of Alzheimer's disease. *CNS & Neurological Disorders-Drug Targets (Formerly Current Drug Targets-CNS & Neurological Disorders)*, 16(10), 1050–1065.

Matsuyama, T., Kubli, S. P., Yoshinaga, S. K., *et al* (2020). An aberrant STAT pathway is central to COVID-19. *Cell Death and Differentiation*, 27(12), 3209–3225.

McDermott, D. F. and Atkins, M. B. (2013). PD-1 as a potential target in cancer therapy. *Cancer Medicine*, 2(5), 662–673.

McGonagle, D., Sharif, K., O'Regan, A., *et al* (2020). The role of cytokines including interleukin-6 in COVID-19 induced pneumonia and macrophage activation syndrome-like disease. *Autoimmunity Reviews*, 19(6), 102537.

Mehta, P., McAuley, D. F., Brown, M., Sanchez, E., *et al* (2020). COVID-19, consider cytokine storm syndromes and immunosuppression. *The Lancet*, 395(10229), 1033–1034.

Mitran, M. I., Nicolae, I., Ene, CD., *et al* (2018). Relationship between gamma-glutamyl transpeptidase activity and inflammatory response in lichen planus. *Revista de Chimie*, 69(3), 739–743.

Miyashita, H., Mikami, T., Chopra, N., *et al* (2020). Do patients with cancer have a poorer prognosis of COVID-19? An experience in New York City. *Annals of Oncology*, 31(8), 1088–1089.

Monaghan, P., Metcalfe, N. B., Torres, R., *et al* (2009). Oxidative stress as a mediator of life history trade-offs, mechanisms, measurements and interpretation. *Ecology Letters*, 12(1), 75–92.

Montopoli, M., Zumerle, S., Vettor, R., *et al* (2020). Androgen-deprivation therapies for prostate cancer and risk of infection by SARS-CoV-2, a population-based study (N= 4532). *Annals of Oncology*, 31(8), 1040–1045.

Moore, P. S. and Chang, Y. (2010). Why do viruses cause cancer? Highlights of the first century of human tumour virology. *Nature Reviews Cancer*, 10(12), 878–889.

Naidu, M. S., Suryakar, A. N., Swami, S. C., *et al* (2007). Oxidative stress and antioxidant status in cervical cancer patients. *Indian Journal of Clinical Biochemistry*, 22(2), 140–144.

Nguyen, M. L., Kraft, R. M., Aubert, M., *et al* (2007). P53 and hTERT determine sensitivity to viral apoptosis. *Journal of Virology*, 81(23), 12985–12995.

Norooznezhad, A. H. and Mansouri, K. (2021). Endothelial cell dysfunction, coagulation, and angiogenesis in coronavirus disease 2019 (COVID-19). *Microvascular Research*, 137, 104188.

Ott, M., Gogvadze, V., Orrenius, S., *et al* (2007). Mitochondria, oxidative stress and cell death. *Apoptosis*, 12(5), 913–922.

Paces, J., Strizova, Z., Daniel, S. M. R. Z., *et al* (2020). COVID-19 and the immune system. *Physiological Research*, 69(3), 379.

Pampin, M., Simonin, Y., Blondel, B., *et al* (2006). Cross talk between PML and p53 during poliovirus infection, implications for antiviral defense. *Journal of Virology*, 80(17), 8582–8592.

Parkin, D. M., Pisani, P., Munoz, N., *et al* (1999). The global health burden of infection associated cancers. *Cancer Surveys*, 33, 5–33.

Passamonti, F., Cattaneo, C., Arcaini, L., *et al* (2020). Clinical characteristics and risk factors associated with COVID-19 severity in patients with haematological malignancies in Italy, a retrospective, multicentre, cohort study. *The Lancet Haematology*, 7(10), e737–e745.

Paul, S., Rausch, C. R., Jain, N., *et al* (2021). Treating leukemia in the time of COVID-19. *Acta Haematologica*, 144(2), 132–145.

Percival, M. E. M., Lynch, R. C., Halpern, A. B., *et al* (2020). Considerations for managing patients with hematologic malignancy during the COVID-19 pandemic, the Seattle strategy. *JCO Oncology Practice*, 16(9), 571–578.

Poprac, P., Jomova, K., Simunkova, M., *et al* (2017). Targeting free radicals in oxidative stress-related human diseases. *Trends in Pharmacological Sciences*, 38(7), 592–607.

Pothuri, B., Secord, A. A., Armstrong, D. K., *et al* (2020). Anti-cancer therapy and clinical trial considerations for gynecologic oncology patients during the COVID-19 pandemic crisis. *Gynecologic Oncology*, 158(1), 16–24.

Ramaiah, M. J. (2020). mTOR inhibition and p53 activation, microRNAs, The possible therapy against pandemic COVID-19. *Gene Reports*, 20, 100765.

Ratre, Y. K., Kahar, N., Bhaskar, L. V. K. S., *et al* (2021). Molecular mechanism, diagnosis, and potential treatment for novel coronavirus (COVID-19), a current literature review and perspective. *3 Biotech*, 11(2), 1–24.

Read, S. A. and Douglas, M. W. (2014). Virus induced inflammation and cancer development. *Cancer Letters*, 345(2), 174–181.

Reuter, S., Gupta, S. C., Chaturvedi, M. M., *et al* (2010). Oxidative stress, inflammation, and cancer, how are they linked? *Free Radical Biology and Medicine*, 49(11), 1603–1616.

Robilotti, E. V., Babady, N. E., Mead, P. A., *et al* (2020). Determinants of COVID-19 disease severity in patients with cancer. *Nature Medicine,* 26, 1218–1223.

Romano, G., Sgambato, A., Mancini, R., *et al* (2000). 8-hydroxy-2′-deoxyguanosine in cervical cells, correlation with grade of dysplasia and human papillomavirus infection. *Carcinogenesis*, 21(6), 1143–1147.

Ruan, Q., Yang, K., Wang, W., *et al* (2020). Clinical predictors of mortality due to COVID-19 based on an analysis of data of 150 patients from Wuhan, China. *Intensive Care Medicine*, 46(5), 846–848.

Rugge, M., Zorzi, M. and Guzzinati, S. (2020). SARS-CoV-2 infection in the Italian Veneto region, adverse outcomes in patients with cancer. *Nature Cancer*, 1(8), 784–788.

Sacconi, A., Donzelli, S., Pulito, C., *et al* (2020). TMPRSS2, a SARS-CoV-2 internalization protease is downregulated in head and neck cancer patients. *Journal of Experimental & Clinical Cancer Research*, 39(1), 1–15.

Salimi, S. and Hamlyn, J. M. (2020). COVID-19 and crosstalk with the hallmarks of aging. *The Journals of Gerontology, Series A*, 75(9), e34–e41.

Scialò, F., Fernández-Ayala, D. J., Sanz, A., *et al* (2017). Role of mitochondrial reverse electron transport in ROS signaling, potential roles in health and disease. *Frontiers in Physiology*, 8, 428.

Sedjo, R. L., Roe, D. J., Abrahamsen, M., *et al* (2002). Vitamin A, carotenoids, and risk of persistent oncogenic human papillomavirus infection. *Cancer Epidemiology Biomarkers and Prevention*, 11(9), 876–884.

Shereen, N. G. and Salman, D. (2019). Delivering chemotherapy at home, how much do we know? *British Journal of Community Nursing*, 24(10), 482–484.

Shibue, T., Suzuki, S., Okamoto, H., *et al* (2006). Differential contribution of Puma and Noxa in dual regulation of p53-mediated apoptotic pathways. *The EMBO Journal*, 25(20), 4952–4962.

Shim, J. H., Kim, K. H., Cho, Y. S., *et al* (2008). Protective effect of oxidative stress in HaCaT keratinocytes expressing E7 oncogene. *Amino Acids*, 34(1), 135–141.

Siegel, E. M., Patel, N., Lu, B., *et al* (2012a). Biomarkers of oxidant load and type-specific clearance of prevalent oncogenic human papillomavirus infection, Markers of immune response? *International Journal of Cancer*, 131(1), 219–228.

Siegel, E. M., Patel, N., Lu, B., *et al* (2012b). Circulating biomarkers of iron storage and clearance of incident human papillomavirus infectioniron status and HPV clearance. *Cancer Epidemiology, Biomarkers and Prevention*, 21(5), 859–865.

Simmons, G., Reeves, J. D., Rennekamp, A. J., *et al* (2004). Characterization of severe acute respiratory syndrome-associated coronavirus (SARS-CoV) spike glycoprotein-mediated viral entry. *Proceedings of the National Academy of Sciences*, 101(12), 4240–4245.

Stewart, C. A., Gay, C. M., Ramkumar, K., *et al* (2021). Lung cancer models reveal SARS-CoV-2-induced EMT contributes to COVID-19 pathophysiology. *BioRxiv*. http://doi.org.10.1101/2020.05.28.122291.

Sungnak, W., Huang, N., Bécavin, C., *et al* (2020). SARS-CoV-2 entry factors are highly expressed in nasal epithelial cells together with innate immune genes. *Nature Medicine*, 26(5), 681–687.

Tan, L., Wang, Q., Zhang, D., *et al* (2020). Lymphopenia predicts disease severity of COVID-19, a descriptive and predictive study. *Signal Transduction and Targeted Therapy*, 5(1), 1–3.

Valko, M., Leibfritz, D., Moncol, J., *et al* (2007). Free radicals and antioxidants in normal physiological functions and human disease. *The International Journal of Biochemistry and Cell Biology*, 39(1), 44–84.

Van de Haar, J., Hoes, L. R., Coles, C. E., *et al* (2020). Caring for patients with cancer in the COVID-19 era. *Nature Medicine*, 26(5), 665–671.

Varadarajan, S., Balaji, T. M., Sarode, S. C., *et al* (2020). EMMPRIN/BASIGIN as a biological modulator of oral cancer and COVID-19 interaction, novel propositions. *Medical Hypotheses*, 143, 110089.

Vargas, A. J. and Harris, C. C. (2016). Biomarker development in the precision medicine era, lung cancer as a case study. *Nature Reviews Cancer*, 16(8), 525–537.

Vuagnat, P., Frelaut, M., Ramtohul, T., *et al* (2020). COVID-19 in breast cancer patients, a cohort at the institute curie hospitals in the Paris area. *Breast Cancer Research*, 22(1), 1–10.

Wang, F., Liu, R., Yang, J., *et al* (2021). New insights into genetic characteristics between multiple myeloma and COVID-19, An integrative bioinformatics analysis of gene expression omnibus microarray and the cancer genome atlas data. *International Journal of Laboratory Hematology*, 43(6), 1325–1333.

Wang, J., Liu, Y., & Zhang, S. (2021). Prognostic and immunological value of ATP6AP1 in breast cancer, implications for SARS-CoV-2. *Aging (Albany NY)*, 13(13), 16904.

Wei, X., Su, J., Yang, K., *et al* (2020). Elevations of serum cancer biomarkers correlate with severity of COVID-19. *Journal of Medical Virology*, 92(10), 2036–2041.

Williams, A. E. and Chambers, R. C. (2014). The mercurial nature of neutrophils, still an enigma in ARDS? *American Journal of Physiology-Lung Cellular and Molecular Physiology*, 306(3), L217–230.

Wu, C., Chen, X., Cai, Y., *et al* (2020). Risk factors associated with acute respiratory distress syndrome and death in patients with coronavirus disease 2019 pneumonia in Wuhan, China. *JAMA Internal Medicine*, 180(7), 934–943.

Wu, Z. and McGoogan, J. M. (2020). Characteristics of and important lessons from the coronavirus disease 2019 (COVID-19) outbreak in China, summary of a report of 72 314 cases from the Chinese center for disease control and prevention. *Jama*, 323(13), 1239–1242.

Yang, F., Shi, S., Zhu, J., *et al* (2020). Clinical characteristics and outcomes of cancer patients with COVID-19. *Journal of Medical Virology*, 92(10), 2067–2073.

Ye, Z. W., Zhang, J., Townsend, D. M., *et al* (2015). Oxidative stress, redox regulation and diseases of cellular differentiation. *Biochimica et Biophysica Acta (BBA)-General Subjects*, 1850(8), 1607–1621.

Yuan, L., Chen, Z., Song, S., *et al* (2015). P53 degradation by a coronavirus papain-like protease suppresses type I interferon signaling. *Journal of Biological Chemistry*, 290(5), 3172–3182.

Yuan, X., Yao, Z., Wu, J., Zhou, Y., *et al* (2007). G1 phase cell cycle arrest induced by SARS-CoV 3a protein via the cyclin D3/pRb pathway. *American Journal of Respiratory Cell and Molecular Biology*, 37(1), 9–19.

Zhang, L., Wei, L., Jiang, D., *et al* (2007). SARS-CoV nucleocapsid protein induced apoptosis of COS-1 mediated by the mitochondrial pathway. *Artificial Cells, Blood Substitutes and Biotechnology*, 35(2), 237–253.

Zhang, L., Zhu, F., Xie, L., *et al* (2020). Clinical characteristics of COVID-19-infected cancer patients, a retrospective case study in three hospitals within Wuhan, China. *Annals of Oncology*, 31(7), 894–901.

Zhang, Y., Mao, Q., Li, Y., *et al* (2021). Cancer and COVID-19 susceptibility and severity, a two-sample mendelian randomization and bioinformatic analysis. *Frontiers in Cell and Developmental Biology*, 9, 759257.

Zong, Z., Wei, Y., Ren, J., *et al* (2021). The intersection of COVID-19 and cancer, signaling pathways and treatment implications. *Molecular Cancer*, 20(1), 1–9.

Zou, X. C., Wang, Q. W. and Zhang, J. M. (2017). Comparison of 5–FU-based and capecitabine-based neoadjuvant chemoradiotherapy in patients with rectal cancer, a meta-analysis. *Clinical Colorectal Cancer*, 16(3), e123–e139.

Zur Hausen, H (1991). Viruses in human cancers. *Science*, 254(5035), 1167–1173.

8 COVID-19 Severity in Lung Cancer Patients and Current Treatment

Saurabh Kumar, Osaid Masood, Shama Parveen, Sanjay Saini and Monisha Banerjee

8.1 INTRODUCTION

The SARS-CoV-2 virus, a new coronavirus, discovered in Wuhan (China) for the first time led to a pandemic, resulting in many people died and complications worldwide. Patients (75% of cases) with coronavirus disease 2019 (COVID-19) are found to have at least one COVID-19 related comorbidity. Hypertension, diabetes, cancer, chronic obstructive pulmonary disease (COPD), endothelial dysfunction and cardiovascular disease are among the most commonly reported comorbidities (Ejaz *et al*, 2020). SARS-CoV-2, the pathogen responsible for COVID-19, spreads through inhaled droplets in the respiratory system. According to various researchers, COVID-19 patients with lung cancer or COPD have a higher mortality rate than those with other cancers. This may be owing to a combination of unique pathophysiological features, such as a history of smoking that has resulted in the deterioration of the lungs and increasing constraints due to the pandemic's impact on respiratory healthcare services. Inflammation in the lungs of COPD cases causes blockage, inadequate airflow and irreversible loss of lung function. Severe pneumonia and poor prognosis are more likely in COPD patients who acquire COVID-19. Patients with COPD are particularly vulnerable to viral respiratory tract infections that can cause respiratory exacerbations. Exacerbation is among the most common causes of COPD is a viral respiratory infection (AECOPD). The underlying susceptibility of COVID-19 to lung tissue necessitates an assessment of the burden of lung-specific comorbidities, such as COPD, lung cancer, asthma and cystic fibrosis, among COVID-19 patients. People with underlying comorbidities need to be aware of the importance of this in order to improve their outcomes.

DOI: 10.1201/9781003324911-12

8.2 IMPACT OF LUNG CANCER AND COPD ON THE SEVERITY OF COVID-19

Multiple surveys have found that individuals with cancer who contracted COVID-19 had significantly worse prognosis, including greater rates of hospitalisation and death. COVID-19 symptoms appeared to be more severe in those who had lung cancer or other pre-existing diseases, such as a history of smoking, genetic variation in immunity, underlying pulmonary disease, and/or cancer-directed therapy. Lung cancer patients were particularly vulnerable to COVID-19 infection because of their advanced age and other medical conditions. Their immunocompromised state as a result of the disease or its treatment, the use of supportive medications such as steroids and the high frequency of contact with healthcare system also resulted in higher susceptibility. There has been consistent evidence that patients with lung cancer have a higher mortality rate than those with other types of cancer.

People with COVID-19 admitted to the hospital were less likely to have COPD than the general population, according to early studies conducted during the pandemic (Leung *et al*, 2020). COPD is a risk factor for hospitalisation with COVID-19, but only modestly so, according to better-designed studies published in late 2021 (Halpin *et al*, 2020). In a cohort of 8.28 million primary care patients, the adjusted hazard ratio for COPD hospitalisation was 1.54 [95% CI 1.45–1.63] (Aveyard *et al*, 2021). COPD was also found to be an independent risk factor for many other diseases or death (Halpin *et al*, 2020; Beltramo *et al*, 2021).

When COVID-19 first emerged in the epidemic, there were concerns about the long-term effects on patients with COPD. It is probable that the maintenance therapy is reducing the likelihood of developing COPD; however, there are also some unexpected findings that could have a substantial impact on the management of the disease in the future (Beltramo *et al*, 2021).

The course of COVID-19 was reported to be longer and more severe in COPD and lung cancer patients than in the general population. Approximately one-third of lung cancer patients in the United States had a milder outpatient course, whereas, two-third required hospitalisation, with a mortality rate of one-quarter. Although these cases were more severe, the observed phenotypes of COVID-19 sickness were similar to those found in the general population, according to the literature (Mohanty *et al*, 2021). Several other factors, such as smoking history, age, hypertension and COPD were found to be associated with a wide range of COVID-19 phenotypes. In a retrospective investigation, out of 72% patients with metastatic or aggressive lung cancer, 62% were hospitalised with a total mortality rate of 25%; however, deaths due to COVID-19 comprised only 11% (Mohanty *et al*, 2021). Another study reported 32% mortality and 72% hospitalisation of 1012 lung cancer patients but a remarkably low ICU admission (12%) and mechanical ventilation (7%) (Aleebrahim-Dehkordi *et al*, 2020). All data show that lung cancer is a significant risk factor for developing SARS-CoV-2 infection. The study designs and medical practices were variable at the time of COVID-19; therefore, the data discussed here are heterogeneous in nature.

COVID-19 has an impact on not only the treatment of lung cancer and COPD patients, but also on numerous aspects of lung cancer and COPD research. As a result of the pandemic, clinical research has been altered. Researchers are now coming up with new and creative ways to carry out cancer research. It is becoming more and more obvious that until people are adequately vaccinated against COVID-19, the incidence of COVID-19 in people with lung cancer will continue to rise. Lung cancer patients and researchers have been adversely impacted by the SARS-CoV-2 epidemic. Lung cancer patients are more likely than the general public to contract the virus and suffer severe outcomes. Further research is required to figure out why lung cancer patients are more susceptible to COVID-19 infection and are prone to early death.

8.3 PREVALENCE, DIAGNOSIS, MANAGEMENT AND PREVENTION OF COVID-19 IN LUNG CANCER OR COPD

8.3.1 Prevalence

As of June 2022, World Health Organization (WHO, 2022) had received reports of 528,816,317 confirmed cases of COVID-19, resulting in 6,294,969 deaths. The severe acute respiratory syndrome coronavirus-2 (SARS-CoV2) epidemic in Wuhan, China in December of 2019 led to a global pandemic of the respiratory tract infection known as COVID-19. In January 2021, there were more than 400,000 deaths linked to COVID-19 in the United States, which was the epicentre of the outbreak. Smoking and COPD were observed to increase the incidence of severe patients with lung cancer who had the COVID-19 condition (Kulkarni *et al*, 2021).

8.3.2 Diagnosis

In addition to coughing up mucus, dyspnea and an occasional fever, COVID-19 patients with lung cancer may also present other symptoms produced by the tumour itself. As a result of therapeutic agents including chemotherapy, radiation and immunotherapy, patients may experience side effects associated with the treatment that might affect the lungs. A wide range of symptoms can be seen in computed tomography scans (CTs) of chest in lung cancer patients. Early identification of COVID-19 is extremely difficult in lung cancer patients, therefore requiring a thorough investigation for differential diagnosis and etiological identification. Chest computed tomography (CT) is the preferable imaging approach for the diagnosis, staging and follow-up of patients with lung cancer. Upto 97% of the time, CT can detect COVID-19 more accurately than RT-PCR during diagnosis of the disease.

There is a possibility that it will be difficult to discern the symptoms of COVID-19 infection and COPD, anosmia, dysgeusia, lethargy, diarrhoea and nausea/vomiting are all common symptoms in COVID-19 patients. Coughing and shortness of breath are also seen in more than 60% of those who have the disease.

In the beginning, the symptoms of COVID-19 may be mild, but lung function can quickly decline. People with COPD who already have a reduced pulmonary reserve may find the prodrome of milder symptoms particularly distressing. It is important to have a high level of suspicion for COVID-19 in patients with COPD who report indicators of an exacerbation, such as fever, foul taste/smell or gastrointestinal difficulties.

Patients with COPD may have persistent symptoms, which may complicate diagnosis. According to a recent study, SARS-CoV-2-positive patients restored to their pre-infection health in only 65% of cases (Tenforde *et al*, 2020). Cough, exhaustion and shortness of breath can last for days, weeks or even months in some people. People with numerous chronic medical illnesses were more likely to experience a delayed recovery, although this was not particularly associated with those having COPD.

8.3.3 Management

During the COVID-19 pandemic, it is critical to keep patients and employees as safe as possible while also dealing with the most life-threatening elements of the disease. This can be accomplished by avoiding unnecessary in-person encounters with medical professionals and tips to the clinic or hospital. SARS-CoV-2 infection should be tested in patients who need to be admitted to the hospital if they exhibit any of the symptoms. COVID-19 testing should be performed whenever possible in patients having any invasive operation or systemic chemotherapy and immunotherapy. During the pandemic, there has been a decrease in overall clinical trial enrolment. Ablative radiation is the treatment of choice for stage I/II and resectable stage III NSCLC (Non-Small Cell Lung Cancer). Surgery for lung cancer is unchanged amid this year's COVID-19 epidemic. The logistics of early-stage lung cancer clinical practise, on the other hand, may be altered.

As a result of the pandemic, questions have been raised about the appropriateness of changing treatment regimens in COPD patients. Since COVID-19 and COPD share many common symptoms, questions have been raised concerning how to distinguish between the two (Halpin *et al*, 2021). Patients with COPD may or may not be more susceptible to COVID-19 infection. It is important to keep in mind that some aspects of managing patients with COPD can be done remotely, which could lessen the risk of infection.

In recent years, the use of telehealth services has grown rapidly. Large-scale screening of patients, distant clinical interactions or professional supervision of patient care is all now possible with the help of telehealth services (Kaye *et al*, 2020). Telehealth improves the flow of healthcare systems by providing greater comfort and a more patient-centric approach (Dockendorf *et al*, 2021). Telemedicine and telehealthcare are two types of telehealth services that can be provided to patients *via* telecommunications technology. Telehomecare, telenursing, telecoaching and telerehabilitation are all types of telehealthcare. Teleradiology, teledermatology, telepsychiatry, telecardiology and telerehabilitation are a few of the teleservices that can be used in conjunction with telemedicine. These services have already been effectively implemented through the use of telemonitoring as well as telerehabilitation. This field's home telehealth services have been shown to lower hospitalisation and emergency room (ER) visit rates. Disease exacerbations are well-tracked using the telerehabilitation programme. Teleconsultations and telespirometry between healthcare professionals (physicians and nurses) and pulmonary specialists for the treatment and care of patients in remote locations have been the driving forces behind this development. Telespirometry uses remote determination of lung capacity to find out if someone has COPD and to see how it is getting worse. Spirometry may only be utilised when it is necessary for the diagnosis of COPD and/or the assessment of status of lung function for treatments or surgery during periods of high COVID-19 prevalence in the population (Franczuk *et al*, 2020). Thus, remote diagnostic approaches appear to be of enormous practical importance.

Recently published studies on telerehabilitation interventions have shown that it is effective to in-hospital Public Relations (PR) programmes with an established safety and feasibility (Taito *et al*, 2021). Telerehabilitation has many advantages, including less stress on caregivers, lower costs, following recommendations, exercise progression, an environment in the home and objective monitoring of physical activity (Rutkowski, 2021). Reductions in dyspnea, functional ability and quality of life were shown to be the most significant outcomes. Using telerehabilitation in the same way as traditional PR helps patients with their shortness of breath, ability to exercise and lung function, While also lowering their risk of hospitalisation and death. Online, face-to-face or phone calls were used to do supervised training utilising performance training equipment (treadmill or cycling) and free exercise (fitness, cardio, strength, yoga) (Gonzalo-Encabo *et al*, 2022). Patients with mild to severe diseases received treatment. Using a virtual trainer or pre-recorded videos, a second group of interventions is possible. Since people with COPD and COVID-19 infection have a higher risk of malnutrition and muscle loss, hospital care should include nutritional support and early mobilisation.

8.3.4 Prevention

If a person has lung cancer and is at risk of contracting the coronavirus, the following preventive actions should be implemented without abandoning the therapy:

- Hands should be thoroughly cleaned for at least 20 seconds on a frequent basis before and after eating, after using the bathroom or blowing nose, coughing and sneezing, before and after coming into contact with any other person, like handshake, *etc*.
- Touching of eyes, nose or mouth should be avoided at all costs.
- Being around contagious people who have flu or cold should be avoided.
- A hand sanitizer containing at least 60% alcohol should be used whenever one has to go out.

- If one is experiencing any of the symptoms, it is best to stay home and avoid social gatherings.
- Regular sanitisation of the entire house or spots that are frequently touched, such as doorbell, door handles, *etc*., should be done.

In order to prevent SARS-CoV-2 infection, patients with COPD should practise basic infection control measures, such as avoiding close contact and washing hands. Wearing a face mask can help prevent the transmission of disease. It is not known if surgical masks and N95 respirators are useful in protecting patients from infection, but both have been shown to be successful in avoiding influenza-like illness and laboratory-confirmed influenza among healthcare personnel (Boškoski *et al*, 2020). It is critical that patients with chronic lung illness wore face masks during the pandemic of COVID-19, according to a statement from the American College of Chest Physicians, American Lung Association, American Thoracic Society (ATS) and COPD Foundation. A tight-fitting N95 mask increases the resistance to inhalation. Having to wear a N95 mask for 10 minutes at rest accompanied by 6 minutes of walking had a negative effect on COPD patients' breathing rate, peripheral oxygen saturation and exhaled CO2 levels, while wearing a surgical mask does not seem to affect breathing, even in patients with severe airflow limitation (Hopkins *et al*, 2021).

Protecting those who are at risk from contacting the coronavirus by shielding or sheltering them can be used as an alternative to more drastic measures such as lockdowns or complete physical separation.

8.4 CHALLENGES FOR ACCESS TO CARE FOR COVID-19 PATIENTS WITH LUNG CANCER OR COPD

Patients with lung cancer are more susceptible to acquiring the severe acute respiratory syndrome coronavirus 2 (SARS-CoV-2) (Rolfo *et al*, 2021). According to Yu *et al* (2020) cancer patients have a higher chance of getting SARS-CoV-2 infection (0.79%) than non-cancer patients (0.37%). A complicated combination of characteristics linked to lung cancer, such as smoking-related lung damage, significant cardiovascular and respiratory comorbidities and advanced age exacerbates SARS-CoV-2 infection. Due to pulmonary and alveolar architectural flaws produced by previous thoracic surgery or radiotherapy and malignant airway obstruction, patients with lung cancer have an increased risk of infection. The Clinical Advisory Group of the UK Lung Cancer Coalition observed that individuals who contracted COVID-19 following lung cancer surgery had an elevated mortality incidence of 40 to 50%. (Passaro *et al*, 2021). Patients should be tested for COVID-19 prior to surgery to limit the risk of surgery during the incubation period (Ribeiro *et al*, 2021).

Changes in tumour microenvironment result from those in the alveolar epithelium and pulmonary arteries, infiltration of immune cells and tissue-resident macrophages, which is critical for increased innate immunity and inflammation

(Qian *et al*, 2020). Cytokine release is more likely because of this extensive immunological component inside the alveolar epithelium. Alveolar damage and fibrin deposits are also risk factors for both small and large vessel pulmonary vascular thrombosis. These two pathogenetic pillars are assumed to constitute the critical processes leading to the SARS-CoV-2 infection; people had severe lung damage and acute respiratory distress syndrome (ARDS) (Domingo *et al*, 2020).

Patients with lung cancer have been an especially vulnerable patient population during the COVID-19 pandemic because of pathological and clinical characteristics, as well as the lack of antineoplastic therapy and concomitant medicines. Patients with lung cancer who have been infected with COVID-19 are also experiencing disruptions in healthcare.

Several real-world datasets were used to assess the risk of cancer patients' hospitalisation, complications and death during the COVID-19 pandemic, and data was mostly acquired retrospectively through multicenter registry-based observational cohort studies all around the world (Garassino *et al*, 2020; Kumar *et al*, 2021).

In the early stages of the pandemic, due to a lack of personal protective equipment, insufficient COVID-19 testing materials and overburdened healthcare systems, lung cancer treatments were delayed or abandoned altogether. COVID-19 infection in cancer patients, as well as delays in cancer treatment, can have a negative impact on patients' cancer outcomes. Patients in the Thoracic Cancers International COVID-19 Collaboration (TERAVOLT) registry waited an average of 21 days for oncologic treatment. Twenty-nine immunocompetent people with COVID-19 should be quarantined for 10 days, whereas immunocompromised people should be quarantined for 20 days, according to the Centre for Disease Control and Prevention (CDC) (Passaro *et al*, 2021). COVID-19 problems may be exacerbated by 46 Immune checkpoint inhibitors (ICIs). The fear was that ICIs could enhance the cytokine storm that causes COVID-19 ARDS, as well as the risk of ICIs and COVID-19 pneumonitis overlapping.

Lung cancer patients (102) treated at Memorial Sloan Kettering Cancer Center, showed that patients treated with ICIs alone, without chemotherapy, had outcomes that were equivalent to or better than those treated with other cancer options. Given these findings, treatment with immunotherapy alone should be highly considered over chemoimmunotherapy for patients with high Programmed Cell Death Ligand 1 (PD-L1) expression (>50%). Immunotherapy regimens that are taken less frequently can and should be evaluated for alternative dosing. This can reduce both patient and provider exposures while maintaining efficacy, allowing the risk-benefit ratio to be optimised. Both the European Medicines Agency (EMA) and the US Food and Drug Administration have approved this (Caffrey and Borrelli, 2020).

Pembrolizumab (400 mg) is administered every six weeks (FDA) (Lala *et al*, 2020). Nivolumab can now be dosed every four weeks instead of every other week. Atezolizumab (50 mg) can be given every four weeks at a dose of 1680 mg (Morrissey *et al*, 2019). Durvalumab (1500 mg) every four weeks were initially approved, and efficacy in the small cell population was proven (Mathieu *et al*, 2021). The every four-week dose schedule can also be used for stage III non-small cell lung cancer (NSCLC) consolidation.

Depending on the clinical setting, pneumonitis seen on CT scans in patients with lung cancer has typically been classified as immunotherapy pneumonitis, radiation pneumonitis or progressive lung cancer (Xu *et al*, 2020). COVID-19 pneumonitis has also been seen in imaging, and it can be difficult to distinguish between the two. CT results suggestive of COVID-19 should prompt an adequate infectious workup for COVID-19 while keeping the rest of the difference in mind. Patients with cancer, particularly lung cancer, should receive an approved COVID-19 vaccination, as recommended by the American Association for Cancer Research (AACR), National Comprehensive Cancer Network (NCCN), the Society for Immunotherapy of Cancer (SITC), American Society of Clinical Oncology (ASCO), European Society for Medical Oncology (ESMO) as well as those undergoing treatments such as chemotherapy, target agents and immunotherapy. Furthermore, as novel medicines to cure or prevent COVID-19 infection are developed, studying the efficacy and side effects of these medications, as well as their interactions with anticancer agents, including vaccination effects, will be critical for continuing the care for lung cancer patients. Antiviral and antibody medications can be used to treat people who are at high risk of severe COVID-19 sickness. People with a weaker immune system, such as cancer patients, fall into this category.

A case study of COVID-19 in a COPD patient was studied in the Iraqi population, as per Surgical Case Report (SCARE) 2020 guidelines (Essa *et al*, 2021). A 70-year-old man was shifted to the COVID-19 Centre at Rania Teaching Hospital in Kurdistan Region of Iraq, complaining of shortness of breath, confusion and sudden unconsciousness. He had COPD for one year and had no previous medical or surgical history. He smoked heavily, with a 58-pack-year history. His general practitioner recommended Zithromax tablets and a salbutamol inhaler. The tracheostomy was performed on the fourth day of the patient's illness, and he was given 2 L/min oxygen on ventilator pressure control ventilation mode with FIO_2 = 50, positive end-expiratory pressure = 8, tidal volume 500 without complications, and his SPO_2 was 92%, with antibiotic ceftriaxone (1 mg) for five days, antipyretic, proton pump inhibitors and mucolytic drugs. A foley catheter and a nasogastric tube (NG) were also implanted on the sixth day. Without oxygen therapy, the patient's oxygen saturation (SPO_2) climbed to 92% after seven days of unconsciousness. On the tenth day, the patient's ventilator was removed. This particular case was one of the unusual COVID-19 COPD cases. COPD is a major risk factor for hospitalisation, ICU stays as well as COVID-19 patient mortality (Beltramo *et al*, 2021). The recent research on hydroxychloroquine found no change in COVID-19 severity outcomes among patients with lung cancer, which is consistent with a recent retrospective study in New York City and a randomised controlled study out of China (Dai *et al*, 2020).

Both diseases lung cancer and COVID-19 require the expertise of radiologists for diagnosis and surveillance. COVID-19 imaging findings should be kept in mind by radiologists when reviewing imaging data for lung cancer patients. Chest x-rays, CT scans, and Positron Emission Tomography (PET) scans are the cornerstones for diagnosing, managing and following up on acute and chronic lung diseases like COVID-19 pneumonia, lobar pneumonia, lung cancer, lung metastasis, radiation, pneumonitis, radiation fibrosis and pulmonary arterial thromboembolism, among others. Due to local facilities and societal attitudes to imaging in the era of COVID-19 pandemic, the use of different radiological modalities for suspected or confirmed COVID-19 patients varies internationally. However, CT scan or chest X-ray has poor sensitivity for the diagnosis of COVID-19 pneumonia, since up to 18% of CT scans or chest X-rays appear normal in the early stages of the disease, whereas this drops to only 3% in severe disease. However, in asymptomatic individuals with a negative PCR test, the CT scan is quite helpful in diagnosing COVID-19 infection (Holborow *et al*, 2020).

Lung cancer surgical interventions are divided into three categories: diagnostic, therapeutic and palliative. Surgical intervention is occasionally required, even when interventional radiology has handled the bulk of diagnostic situations. Routine and elective surgical procedures should not be prioritised during the COVID-19 outbreak to prevent the virus from spreading and to protect patients and healthcare staff. Prior to any surgical intervention, all patients should have COVID-19 screening or testing for at least three days and be checked for any symptoms on admission, with a second screening test, if necessary. It should be included in the patient's signed consent to the procedure.

8.5 CONCLUSION

The SARS-CoV-2 pandemic has hurt people with lung cancer and slowed the progress of research into lung cancer. Patients with lung cancer are more likely to get infected with the virus and have a higher chance of getting sick or dying than the rest of the population. At the start of the pandemic, patients with lung cancer were not given as much attention when they were admitted to the ICU or any mechanical ventilation. This was confirmed by the TERAVOLT statistics, which showed that their death rate went up. So far, based on what we know about how COVID-19 affects cancer patients, this strategy does not seem to make sense. This confirms that patients with lung cancer should not be kept out of ICU beds and should be given priority to get fast and complete care, just like other patients.

Even though the epidemic has made things hard, there has been a huge international effort to find out how COVID-19 affects the patient population. During this health crisis, cancer societies like ESMO and ASCO have put out statements and guidelines to help doctors take care of lung cancer patients. In the context of COVID-19, studies like the TERAVOLT analysis and Cancer and COVID-19 Consortium have given important information about how different treatments affect people with lung cancer. With the goal of improving outcomes for people with lung cancer who have been infected with the virus, these analyses give specific advice on the risks of different treatment plans.

ACKNOWLEDGEMENT

SK, SP and OM are grateful to ICMR, CSIR and UGC New Delhi, India for their respective research fellowships. Authors also acknowledge the Centre of Excellence scheme, Government of Uttar Pradesh, Lucknow, DST, DBT, ICMR, UGC, New Delhi, India for all the research facilities at Molecular and Human Genetics Lab, Department of Zoology, University of Lucknow, Lucknow.

REFERENCES

Aleebrahim-Dehkordi, E., Deravi, N., Reyhanian, A., et al. (2020). Chronic non-communicable diseases in the epidemic (COVID-19): Investigation of risk factors, control and care. *Przeglad Epidemiologiczny*, 74(3), 449–456.

Aveyard, P., Gao, M., Lindson, N., et al. (2021). Association between pre-existing respiratory disease and its treatment, and severe COVID-19: A population cohort study. *The Lancet Respiratory Medicine*, 9(8), 909–923.

Beltramo, G., Cottenet, J., Mariet, A. S., et al. (2021). Chronic respiratory diseases are predictors of severe outcome in COVID-19 hospitalised patients: A nationwide study. *European Respiratory Journal*, 58(61😊, 2004474.

Boškoski, I., Gallo, C., Wallace, M. B., et al. (2020). COVID-19 pandemic and personal protective equipment shortage: Protective efficacy comparing masks and scientific methods for respirator reuse. *Gastrointestinal Endoscopy*, 92(3), 519–523.

Caffrey, A. R., & Borrelli, E. P. (2020). The art and science of drug titration. *Therapeutic Advances in Drug Safety*, 11, 2042098620958910.

Dai, M. Y., Chen, Z., Leng, Y., et al. (2020). Patients with lung cancer have high susceptibility of COVID-19: A retrospective study in Wuhan, China. *Cancer Control*, 27(1), 1073274820960467.

Dockendorf, M. F., Hansen, B. J., Bateman, K. P., et al. (2021). Digitally enabled, patient-centric clinical trials: Shifting the drug development paradigm. *Clinical and Translational Science*, 14(2), 445–459.

Domingo, P., Mur, I., Pomar, V., et al. (2020). The four horsemen of a viral Apocalypse: The pathogenesis of SARS-CoV-2 infection (COVID-19). *EBioMedicine*, 58, 102887.

Ejaz, H., Alsrhani, A., Zafar, A., et al. (2020). COVID-19 and comorbidities: Deleterious impact on infected patients. *Journal of Infection and Public Health*, 13(12), 1833–1839.

Essa, R. A., Ahmed, S. K., Bapir, D. H., et al. (2021). Challenge of surviving COPD with COVID-19 patient: Review of the literature with unusual case report. *IJS Global Health*, 4(6), e65.

Franczuk, M., Przybyłowski, T., Czajkowska-Malinowska, M., et al. (2020). Spirometry during the SARS-CoV-2 pandemic. Guidelines and practical advice from the expert panel of respiratory physiopathology assembly of polish respiratory society. *Advances in Respiratory Medicine*, 88(6), 640–650.

Garassino, M. C., Whisenant, J. G., Huang, L. C., et al. (2020). COVID-19 in patients with thoracic malignancies (TERAVOLT): First results of an international, registry-based, cohort study. *The Lancet Oncology*, 21(7), 914–922.

Gonzalo-Encabo, P., Wilson, R. L., Kang, D. W., et al. (2022). Exercise oncology during and beyond the COVID-19 pandemic: Are virtually supervised exercise interventions a sustainable alternative? *Critical Reviews in Oncology/Hematology*, 103699.

Halpin, D. M., Faner, R., Sibila, O., et al. (2020). Do chronic respiratory diseases or their treatment affect the risk of SARS-CoV-2 infection? *The Lancet Respiratory Medicine*, 8(5), 436–438.

Halpin, S. J., McIvor, C., Whyatt, G., et al. (2021). Postdischarge symptoms and rehabilitation needs in survivors of COVID-19 infection: A cross-sectional evaluation. *Journal of Medical Virology*, 93(2), 1013–1022.

Holborow, A., Asad, H., Porter, L., et al. (2020). The clinical sensitivity of a single SARS-CoV-2 upper respiratory tract RT-PCR test for diagnosing COVID-19 using convalescent antibody as a comparator. *Clinical Medicine*, 20(6), e209.

Hopkins, S. R., Dominelli, P. B., Davis, C. K., et al. (2021). Face masks and the cardiorespiratory response to physical activity in health and disease. *Annals of the American Thoracic Society*, 18(3), 399–407.

Kaye, R., Rosen-Zvi, M., & Ron, R. (2020). Digitally-enabled remote care for cancer patients: Here to stay. *Seminars in Oncology Nursing*, 36(6), 151091.

Kulkarni, A. A., Wilson, G., Fujioka, N., et al. (2021). Mortality from COVID-19 in patients with lung cancer. *Journal of Cancer Metastasis and Treatment*, 7, 31.

Kumar, G., Mukherjee, A., Sharma, R. K., et al. (2021). Clinical profile of hospitalized COVID-19 patients in first & second wave of the pandemic: Insights from an Indian registry based observational study. *The Indian Journal of Medical Research*, 153(5–6), 619.

Lala, M., Li, T. R., de Alwis, D. P., et al. (2020). A six-weekly dosing schedule for pembrolizumab in patients with cancer based on evaluation using modelling and simulation. *European Journal of Cancer*, 131, 68–75.

Leung, J. M., Niikura, M., Yang, C. W. T., et al. (2020). Covid-19 and COPD. *European Respiratory Journal*, 56(2); 2002108.

Mathieu, L., Shah, S., Pai-Scherf, L., et al. (2021). FDA approval summary: Atezolizumab and durvalumab in combination with platinum-based chemotherapy in extensive stage small cell lung cancer. *The Oncologist*, 26(5), 433–438.

Mohanty, A., Agnihotri, S., Mehta, A., et al. (2021). COVID-19 and cancer: Sailing through the tides. *Pathology-Research and Practice*, 221, 153417.

Morrissey, K. M., Marchand, M., Patel, H., et al. (2019). Alternative dosing regimens for atezolizumab: An example of model-informed drug development in the post marketing setting. *Cancer Chemotherapy and Pharmacology*, 84(6), 1257–1267.

Passaro, A., Bestvina, C., Velez, M. V., et al. (2021). Severity of COVID-19 in patients with lung cancer: Evidence and challenges. *Journal for Immunotherapy of Cancer*, 9(3), e002266.

Qian, J., Olbrecht, S., Boeckx, B., et al. (2020). A pan-cancer blueprint of the heterogeneous tumor microenvironment revealed by single-cell profiling. *Cell Research*, 30(9), 745–762.

Ribeiro, R., Wainstein, A. J. A., de Castro Ribeiro, H. S., et al. (2021). Perioperative cancer care in the context of limited resources during the COVID-19 pandemic: Brazilian society of surgical oncology recommendations. *Annals of Surgical Oncology*, 28(3), 1289–1297.

Rolfo, C., Meshulami, N., Russo, A., et al. (2021). Lung cancer and severe acute respiratory syndrome coronavirus 2 infection: Identifying important knowledge gaps for investigation. *Journal of Thoracic Oncology*, 17(2), 214–227.

Rutkowski, S. (2021). Management challenges in chronic obstructive pulmonary disease in the COVID-19 pandemic: Telehealth and virtual reality. *Journal of Clinical Medicine*, 10(6), 1261.

Taito, S., Yamauchi, K., & Kataoka, Y. (2021). Telerehabilitation in subjects with respiratory disease: A scoping review. *Respiratory Care*, 66(4), 686–698.

Tenforde, M. W., Kim, S. S., Lindsell, C. J., et al. (2020). Symptom duration and risk factors for delayed return to usual health among outpatients with COVID-19 in a multistate health care systems network—United States, March-June 2020. *MMWR Morbidity and Mortality Weekly Report*, 69, 993–998.

World Health Organization. (2022). COVID-19 weekly epidemiological update. *Edition*, 84, March 22, 2022.

Xu, B., Xing, Y., Peng, J., et al. (2020). Chest CT for detecting COVID-19: A systematic review and meta-analysis of diagnostic accuracy. *European Radiology*, 30, 5720–5727.

Yu, J., Ouyang, W., Chua, M. L., & Xie, C. (2020). SARS-CoV-2 transmission in patients with cancer at a tertiary care hospital in Wuhan, China. *JAMA Oncology*, 6(7), 1108–1110.

Part V

Current Therapeutic Approaches

9 COVID-19 Therapy
Molecular Mechanisms, Pharmacological Interventions and Therapeutic Targets

Ruchi Chawla, Krishan Kumar, Mohini Mishra and Varsha Rani

9.1 INTRODUCTION

Coronaviruses are sense-strand or positive-sense RNA viruses which are enveloped with club-shaped spikes all around their surface. [1,2] The novel CoVs-2019 have evolved as beta coronaviruses from severe acute respiratory syndrome CoV (discovered in China, in 2003) and Middle East respiratory syndrome CoV (discovered in Saudi Arabia in 2012). [3,4] Based on similarity with 2003 SARS-CoV, the CoVs-2019 have been termed as SARS-CoV-2/COVID-19. The similarity has been observed in their structure and mechanism of entry into the cells which occurs by binding to angiotensin-converting enzyme 2 (ACE2) receptors via variable receptor-binding domain (RBD) present on the cell surface. [4,5] These CoVs have caused severe infection in human populations and have been difficult to handle. The latest outbreak of CoVs in December 2019, that is, COVID-19 with its epicentre in the seafood market of Wuhan city, Hubei Province of China, turned into a pandemic by rapidly spreading to all parts of the globe, transcending national and international borders within no time. [6] Coronavirus has infected more than fifty crore people, besides taking toll of six lakh people. The scientists have identified seven human coronaviruses (HCoVs): α-type HCoV-229E and HCoV-NL63; the β-type HCoV-HKU1, SARS-CoV, MERS-CoV, HCoV-OC43 and SARS-CoV-2. According to their pathogenicity, HCoVs have been divided into highly pathogenic CoVs: SARS-CoV, MERS-CoV and SARS-CoV-2; mildly pathogenic: HCoVs (HCoV-229E, HCoV-NL63, HCoV-OC43 and HCoV-HKU). [7] As the pathogenesis is complex, the infection from coronavirus might result in both asymptomatic and symptomatic conditions. In some patients generalised symptoms like muscle soreness, cough and mild fever may develop; however, others may develop acute respiratory distress syndrome (ARDS) and multiple organ failure, promptly causing death within a short period of time. [8] Infection of the

DOI: 10.1201/9781003324911-14

lower respiratory tract precipitates in the form of severe pneumonia, occasionally leading to fatal acute lung injury (ALI) and ARDS, which is caused by highly pathogenic HCoVs. High degree of morbidity and mortality from HCoVs is a major issue concerning public health. In comparison, mildly pathogenic HCoVs restrict themselves in the upper respiratory tract, causing cold-like symptoms in healthy individuals. [9–11]

It was observed during the outbreak that treatment of CoVs posed a dual challenge because of rapidly changing genetic makeup of virus and also because of the multiple pathogenic mechanisms by which infection is caused. [1] In the inception, treatment was primarily symptomatic as unfortunately, there were no recommended drugs or vaccines which could result in improved outcomes in patients especially with either suspected or confirmed infection. [12] There was restlessness in the medical fraternity globally to discover therapeutic agents for specific treatment of infection caused by coronaviruses. [13]

In this chapter, genetic structure of the virus has been discussed along with possible target sites which can be exploited for as potential treatment options.

9.2 GENETIC STRUCTURE AND PATHOGENESIS OF CoVs

The term "coronavirus" derives its origin from the Latin term which means "crown" or "wreath". The structure of the virus resembles that of solar corona with a characteristic appearance of virions enveloped with a lipidic layer decorated with club-shaped fringe-like projections called spikes as shown in **Figure 9.1**. [14] Biologically, it belongs to Coronaviridae family (subfamily: Orthocoronavirinae; order: Nidovirales) and have been sub-categorised as alpha, beta, gamma and delta CoVs. Nidovirales exhibit characteristic 5' capped, enveloped, non-segmented, single-strand positive-sense RNA which have extraordinarily large genetic structure of 30 kb with size in the range of 80 to 120 nm and consist of 29,903 nucleotides, 38% of which are glycine and cysteine. [15] The spikes play a unique role in the replication of these viruses. [16]

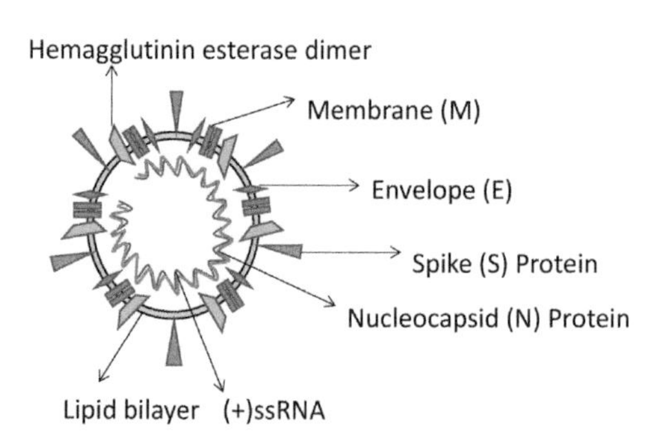

FIGURE 9.1 Structure of SARS-COVID-2.

The genomic set up of CoVs consist of open reading frames (ORFs) (six to ten in number) which encode the replicase proteins, in addition to spike-(S), membrane-(M), envelope-(E) and nucleocapsid-(N) proteins, which are the main structural proteins: The 3′ terminal of the viral genome encodes the ORFs and structural proteins. The access of virus to the host system occurs via the spike proteins; the membrane and envelop proteins assimilate the viral envelope, consisting of a bilayer of lipids and a nucleocapsid having the RNA genome. Further, a glycoprotein known as hemagglutinin-esterase (HE) of approximately 60–65 kDa, is present in most betacoronaviruses. HE trims the acetyl groups off the O-acetylated sialic acid and plays role in binding or release of virus from target site. However, the more precise function of HE still needs to be deciphered. [17,18]

A peculiar condition known as cytokine storm occurs during infection, wherein abnormal release of proinflammatory cytokines occurs, causing acute inflammation at the site of infection, and gradually these cytokines spread to other parts of body. [19] Acute lung injury, acute respiratory distress syndrome (ARDS) and morbidity in SARS-CoV-2 can be credited to aberrant infiltration of inflammatory cells and cytokine storm. [20] Elevated levels of various types of interleukins (IL) like IL-7, IL-8, IL-9, IL-10, IL-1β, IL1-receptor antagonist, basic fibroblast growth factor, granulocyte-macrophage colony-stimulating factor, granulocyte colony-stimulating factor, Interferon gamma-induced protein 10, macrophage inflammatory protein 1-alpha (MIP-1α), MIP-1β, monocyte chemo attractant protein-1, platelet-derived growth factor, tumour necrosis factor -α, vascular endothelial growth factor, and interferon-γ have been found in infected patients; however, the levels of IL-15, IL12p70, IL-5, eotaxin, and RANTES are normal. [21]

9.3 STRUCTURE BASED DRUG TARGETS OF SARS-CoV-2

To understand the drug targets, we need to comprehend the morphology as well as the genomic makeup of SARS-CoV-2. Attachment and access of these viruses to the target structure occurs via receptor-binding domains (RBD) present in the S1 region of spike protein which then attaches to the angiotensin-converting enzyme 2 (ACE2) receptor. During this process, priming of S protein occurs by transmembrane serine protease 2 (TMPRSS2) which cleaves the S protein at S1/S2 and S2' sites following which fusion of viral and host cell membrane occurs and endosomes are formed for entry into the host. [22] Upon entry into the cells, translation ensues forming viral polyproteins which encode the viral replicase complex. Both genomic and sub-genomic RNA are synthesised via its RNA-dependent RNA polymerase. Sub-genomic RNAs generate structural and accessory genes which facilitate assembly and release of viral particles. [23–25] Promising therapeutic targets for treatment of coronavirus can be identified from its structure which interact with host cell as shown in **Figure 9.2**. Drugs based on virus-based targets such as spike glycoprotein (structural protein), 3-chymotrypsin-like protease,

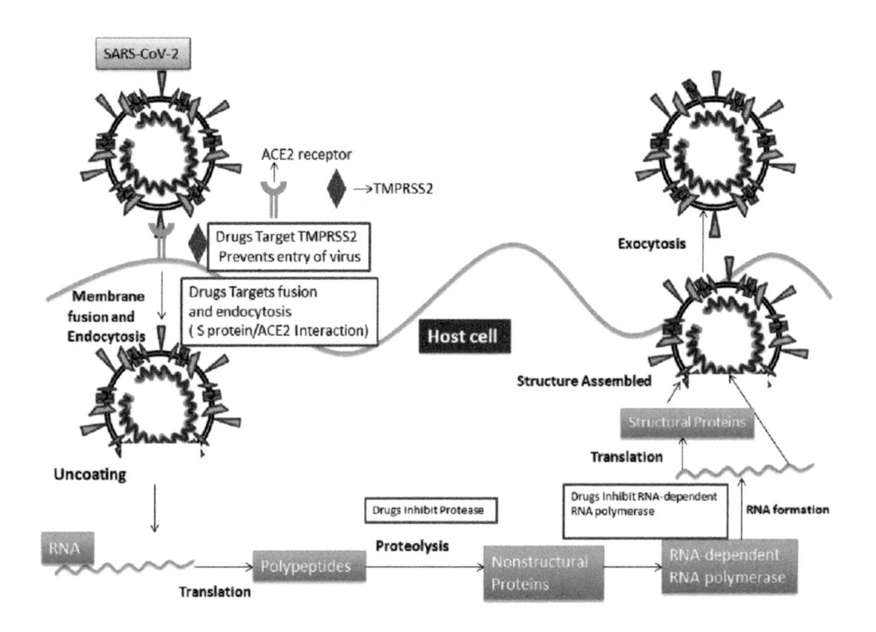

FIGURE 9.2 Life cycle of COVID-19 and potential drug targets.

papain-like protease, helicase and RNA-dependent RNA polymerase (non-structural proteins), and accessory proteins may be used. [18,19] Similarly, in humans (host) endocytosis, endosomal acidification, kinase signalling pathways, angiotensin-converting enzyme 2, etc., can be targeted. [3] Figure 9.2 summarises the therapeutic targets for use for treatment or adjunctive therapy of SARS-CoV-2.

9.4 POTENTIAL TREATMENT OPTIONS OF SARS-CoV-2

9.4.1 Convalescent Plasma Therapy

It is a type of immediate short span immunisation which started off in 1880s in order to treat many viral and bacterial infections. [26] Initially, immunised animals (rabbits and horses) were used as a source of antibodies, which were later obtained from human blood (of recovered individuals). In the year 1890, Von Behring and Kitasato rationally explored the potential of convalescent therapy for the treatment of diphtheria using blood serum. [26–28] History governs the application of convalescent therapy for control of therapeutic conditions like measles, scarlet fever, chickenpox, influenza, Argentine hemorrhagic fever and parvovirus B19 [26,29–33]. This therapy gained importance during the outbreak of Spanish influenza pandemic in 1918. [26,27,34] Similarly, during the outbreak of Western African Ebola virus, MERS-CoV, avian flu, influenza A virus subtypes H1N1 and H5N1 and severe acute respiratory infections, convalescent therapy was found to be promising. [26,30,35–38]

In convalescent therapy, plasma from recovered patients with humoral immunity against infection is transferred to the individuals suffering from same infection, which therein artificially induces passive immunity. [6] Whole blood, plasma, high-titre human Ig, pooled human immunoglobulin, and polyclonal or monoclonal antibodies are different types of convalescent blood products that are used in this therapy; however, plasma is most widely used owing to ease of collection of larger volumes (that too more than once (more frequency of donations)), without causing any adverse effect to the donor. The red blood cells can be infused back into the donor's body. [26,27] However, certain pre-requisites as suggested by WHO (21 CFR 630.10, 21 CFR 630.15) need to be met by the donors, before plasmapheresis can be carried out. [39] Corona negative asymptomatic human subjects aged between 18 to 65 years are considered as apt donors for convalescent therapy. Subjects belonging to regions where diseases like malaria is endemic cannot donate blood. The commonly employed procedure for collection of plasma is apheresis, involving incessant centrifugation of the donor's blood. [27,40,41] Patients were administered with convalescent plasma by Frame et al., on, before and after the tenth day of onset of symptoms of Lassa fever at Jos and Vom hospital, Nigeria. Rapid response to therapy was observed for the patients who received plasma early during the therapy, with a higher rate of survival. [42] Efficacy of this therapy in SARS was evaluated by Cheng et al., on 80 patients suffering from SARS at Prince of Wales Hospital, Hong Kong, for a period of approximately two months (March 20 to May 26, 2003). The convalescent plasma was given after onset of symptoms (around [14]th day), and better outcomes were observed in patients who received plasma before Day 14 as compared to the ones who received it later. A lower mortality rate of 12.5% was observed in patients (n = 80) who received convalescent plasma in comparison to the patients who did not receive plasma (17% mortality (n = 1755)). [35] Shen and his research group transfused convalescent plasma to COVID-19 positive patients (n = 5) with acute respiratory distress syndrome and reported normalisation in the body temperature of four out of five patients within three days, with decrease in the viral load. Moreover, the patients tested negative within twelve days of the therapy. [43] A research group from Korea also reported positive results of convalescent plasma therapy in COVID-19 patients. [44] These case studies validate the use of convalescent plasma therapy against coronavirus and warrant its use in coronavirus infection.

9.4.2 Monoclonal Antibody-Based Therapy

The advancements in the field of molecular biology have led to tremendous developments in monoclonal antibody-based (mAb) therapies which have found multifarious applications in immunotherapeutics and immunological assays. [45] mAbs get bound to the epitope present on the antigen and are produced by B-lymphocyte clone via hybridoma technology. [6,45,46] Since the generation of first hybridoma in 1975, this technology has come a long way with the development of human monoclonal antibodies and Velocimmune humanised mice by

Regeneron Pharmaceuticals, Inc. [47] Due to its high specificity and flexibility, it is being widely explored for diagnostic purposes and targeted therapy for the treatment of various diseases. In the era of personalised medicines, this technique has emerged as a potential approach for treatment of autoimmune diseases, diabetes, cancer, inflammatory and infectious diseases, etc. [45,46] In 1986, the first FDA-approved monoclonal antibody muromonab CD3 was used for kidney transplantation. [48] Monoclonal antibody technique has also been used as a curative treatment for pathogenic coronaviruses like SARS-CoV and MERS-CoV. [49] Further, this technique is also being investigated for SARS-CoV-2. Proteins (membrane-, envelope-, nucleocapsid- and spike-) mediate entry of virus and its replication in the host. As in the case of MERS-CoV, dipeptidyl peptidase 4 acts as target receptor protein and is reported to be responsible for infection. [6,49] Hence, administration of mAbs could effectively inhibit the fusion of RBD and target receptors which can in turn effectively hinder the entry and survival of virus inside the host cells. mAbs such as 80R, CR3014, CR3022, F26G18, F26G19 are being investigated for neutralising SARS-CoV, and those for neutralising MERS-CoV are MERS4, MERS27 4C2, m336 and G4. [49]

9.4.3 Antibody-Based Treatment of SARS-CoV-2

Progressive results have been obtained in the clinical use of monoclonal antibodies in SARS-CoV-2. Monocytes, macrophages and dendritic cells, are activated by respiratory epithelial cells in infection, resulting in secretion of proinflammatory cytokines, interleukin-6 (IL-6). Both circulating and soluble IL-6 receptor complexes indirectly activate endothelial cells, resulting in systemic cytokine storm causing hypotension and acute respiratory distress syndrome (ARDS). Antibodies Sarilumab (Kevzara®, Sanofi, USA, and Regeneron Pharmaceuticals, Inc., USA) and Tocilizumab (Actemra/RoActemra®, F. Hochmann-La Roche AG, Basel, Switzerland) hinder the signal transduction of IL-16 and have shown promising results in COVID-19.[50] Canakinumab (ACZ885, Ilaris, Novartis) a human anti-IL-1β monoclonal antibody, which is generally used for autoimmune disorder, can be used to deal with COVID-19 triggered pneumonia. [51] Novartis initiated a phase 3 human trial of this antibody in patients with COVID-19 pneumonia so as to evaluate the impact on the death rate of patients in the absence of invasive mechanical ventilation. [52] The results of the trial submitted by Novartis were not very supportive and the primary endpoints of the trial did not meet the expected ones, which is that treatment with canakinumab would result in a better chance of survival of the infected individuals with supportive invasive mechanical ventilation up to Day 29. The trial didn't come up with secondary end points as well of reduced COVID-19 induced death rate, and the trial was reported to have failed by Novartis.

Scarce availability of human plasma (from recovered patients) poses limitation to this method of inducing immunity against infection. So the scientists are exploring alternative strategies to generate antibodies against the SARS-CoV-2 viral antigens from non-human sources. Scientists at The Rosalind Franklin

TABLE 9.1
List of Antibodies Tested against SARS-Cov-2 [7], [8]

Antibodies	Clonality and Isotype	Protein structure
SARS-CoV-2 Spike S1 Antibody, SARS-CoV-2 Spike S2 Antibody	Polyclonal	Spike Protein
SARS Matrix Antibody	Polyclonal	Membrane Protein
SARS-CoV-2 Spike S1 recombinant protein	Fc, IgG, His tag, His-Avi tag	Spike S1 Protein Recombinant protein
SARS-CoV-2 Spike-RBD Antibody	Monoclonal	Spike-RBD Protein
SARS-CoV-2 Nucleocapsid Antibody	Polyclonal, Monoclonal	Nucleocapsid Protein
SARS-CoV-2 Nucleoprotein Antibody	Polyclonal	Nucleoprotein
SARS-CoV-2 Envelope Antibody	Polyclonal, Monoclonal	Envelope Protein
SARS-CoV-2 Membrane Antibody	Polyclonal IgG	Membrane protein
SARS-CoV-2 pp1ab Antibody	Polyclonal IgG	pp1ab protein
SARS-CoV-2 ORF10 Antibody	Polyclonal IgG	ORF10Protein
SARS-CoV-2 ORF7a Antibody, SARS-CoV-2 ORF6 Antibody,	Polyclonal IgG	ORF7a, ORF6 Protein

Institute in the UK have isolated two small, stable antibody variants from llamas, South American mammals and found them to be active against coronavirus *in vitro*. [53] Upregulation of granulocytes and macrophage colony-stimulating factor receptor alpha (GM-CSFRα) contributes to severe and acute lung complications in COVID-19 induced pneumonia via induction of growth and development of dendritic cells, monocytes, macrophages and granulocytes. [54,55] Kiniksa Pharmaceuticals commenced clinical trials of Mavrilimumab, a monoclonal antibody that targets GM-CSFRα in severe pneumonia and hyper inflammation in SARS-Cov-2. [56,57] But unfortunately, the results posted by Kiniksa in December 2021 showed that the trial was unable to meet the primary endpoints of proportion (increased survival rate with no requirement of mechanical ventilation at the ^{29}th day of treatment). [56] Some of the SARS-CoV-2 virus antibodies have been mentioned in Table 9.1.

9.4.4 Targeting of Alternative Pathways for the Treatment of SARS-CoV-2

Studies have also suggested that a cascade of Janus/kinase/signal transduction pathway (JAK-STAT) is associated with SARS-CoV-2 and can be targeted for the treatment of infection. Ruxolitinib (Jakafi®, Incyte Corporation, Wilmington, USA), a JAK1/JAK2 inhibitor was assessed for multifocal interstitial pneumonia in a phase 3 study (RUX-COVID (NCT04331665)). According to the study, cytokine storm causes lung inflammation and fluid buildup in the lungs leading to lung damage and breathing problems. So, if Ruxolitinib is given during early stage of the disease, excess production of cytokines can be prevented, but this clinical trial was unable to meet its primary endpoints hence the trial was stopped. [58]

9.5 CURRENTLY USED THERAPEUTIC STRATEGIES FOR TREATMENT SARS-CoV-2

9.5.1 Treatment Based on Drugs

The infection from coronaviruses has plagued the human world time and again during the preceding two decades but there is still lack of any effective treatment for these infections. [35,37] Of the various strategies used for treatment, existing drugs can be repurposed (chosen from the class of anti-malarial drugs, anti-viral drugs and anti-inflammatory drugs) and can fetch positive outcomes in SARS-CoV-2, providing curative as well as symptomatic relief. [59,60] A patient (54-year-old male) administered with Lopinavir/Ritonavir showed recovery from infection with SARS-CoV-2. [61] Methylprednisolone, moxifloxacin, thalidomide, oseltamivir, arbidol hydrochloride, human gamma globulin have also been tried in coronavirus patients. [62] Caly et al., evaluated the *in vitro* efficacy of anti-parasitic drug Ivermectin and reported a ~5000-fold decrease in the viral RNA load after 48 hours of infection. [63] Park et al., in their study demonstrated the efficacy of FDA-approved drugs (lopinavir-ritonavir, hydroxychloroquine sulfate and emtricitabine-tenofovir) against ferret infection model of SARS-CoV-2. Emtricitabine-tenofovir showed lowest viral titres in nasal washes on eighth day post-infection. [64]

9.5.2 Anti-Malarial Drugs

The anti-malarial drugs: chloroquine and its hydroxyl derivative, that is, hydroxychloroquine (though less toxic than chloroquine but exhibits similar therapeutic activity) have been repurposed for malaria, rheumatoid arthritis, lupus erythematosus, etc. Further, these drugs have also been used against different strains of RNA virus (hepatitis A and C, chikungunya, rabies, polio, Lassa, Zika, influenza A and B, influenza A H5N1, HIV, Ebola, Nipah) and DNA virus (herpes simplex and hepatitis B), etc. [60,65] Keyaerts et al., demonstrated the effectiveness of chloroquine against coronavirus OC43 (a human coronavirus) in newborn mice. [66] Suppression of viral replication (human coronavirus 229E) in L132 human epithelial lung cells has been reported upon administration of chloroquine. [50,51] The antiviral activity of chloroquine against SARS-CoV [67] and MERS [68] has also been studied, though activity against MERS still needs to be verified. [69] Significant evidence of its effective use in COVID-19 has been reported. A research group explored the *in vitro* inhibition efficiency of penciclovir, ribavirin, nafamostat, chloroquine, nitazoxanide, favipiravir and remdesivir in clinical isolates of SARS-CoV-2. The results showed effective inhibition of viral infection by chloroquine and remdesivir at low micromolar concentration with high selectivity index as compared to all other drugs used in the study. [3] The dose of chloroquine which inhibits SARS-Cov-2 is comparable to that used in malaria. The half-maximal inhibitory concentration (IC50) of chloroquine used to restrict SARS-CoV-2 replication is 1–3μmol/L. [70]

Various *in vitro* studies have revealed that therapeutic efficacy of hydroxychloroquine is higher than that of chloroquine. [71] Hydroxychloroquine along with azithromycin have shown negative PCR outcome in COVID-19 patients with a recovery period of three to six days. [72] Indian Council of Medical Research (ICMR) approved chemoprophylactic use of hydroxychloroquine in SARS-CoV-2 patients at a dose of 400 mg two times a day, followed by same dose once a week. [73] Hydroxychloroquine and chloroquine, however, show adverse effects such as gastrointestinal upset, retinopathy, abnormal heart rhythms and ventricular tachycardia. The risk might increase when hydroxychloroquine and azithromycin are given together and can lead to heart and kidney complications. [74–76]

The binding of viral protein to ACE2 receptor is blocked by hydroxychloroquine and chloroquine *via* inhibiting terminal glycosylation and thus viral replication. Additionally, these drugs increase pH of the intracellular compartment of endosomes and lysosomes, which affects the integrity of viral genome. [75] The French government had banned antimalarial drugs mainly hydroxychloroquine for the symptomatic treatment of SARS-CoV-2 due to associated cardiac risks. [77]

9.5.3 Antiviral Drugs

9.5.3.1 Remdesivir

The efficacy of anti-viral drug remdesivir to shorten the recovery period in SARS-CoV-2 infected people has been approved by US Food and Drug Administration. After a two-day clinical investigation conducted by the National Institute of Health, intravenously administered remdesivir showed deleterious effect on SARS-CoV-2 contagion. [78] Actually, remdesivir is a drug used in infection caused by Ebola hemorrhagic virus (EBV) and is also found to be effective against SARS-CoV-2 enzymes that inhibit replication of virus. Researchers at Götte's lab have found GS-441524, an adenosine nucleotide analogue, which is a metabolite of remdesivir, and it reduces production of viral RNA by interfering with RNA-dependent RNA polymerase.

Similarly, Gilead Sciences Laboratory also initiated extensive research in SARS-CoV-2 animal models. [79] Chloroquine and remdesivir tested on *in vitro* clinical isolates of SARS-CoV-2 have shown positive outcomes. [3]

9.5.3.2 Lopinavir and Ritonavir

FDA-approved combination therapy of antiretroviral agents Lopinavir and Ritonavir (LPV/RTV) (Kaletra), with protease inhibition activity, which is used for treatment of HIV-positive patients. Clinical studies have been conducted on these drugs to evaluate their application in SARS-CoV-2 infection. Hospitalised adult patients with infection were involved in a controlled and randomised open-label trial and were given standard care of 94% oxygen saturation or less while the patients were taking in atmospheric air or Carrico index was less than 300 mm Hg. The patients were divided randomly into two groups with equal number of patients and were given either standard care alone or a combination of standard care and 400mg lopinavir and 100mg ritonavir, two times a day for 14 days. Administration of lopinavir-ritonavir, however, did not show enhanced effect beyond standard care. [80] Similarly, a 14-day randomised phase 2 open lab trial in adult SARS-CoV-2 patients from six hospitals in Hong Kong was conducted. The patents were allocated arbitrarily in 2:1 ratio and were either given a combination therapy for 14 days consisting of 400 mg lopinavir and 100 mg ritonavir every 12 h, 400 mg ribavirin every 12 h and three doses of 8 million IU interferon beta-1b on alternate days (combination group) or were administered with 400 mg lopinavir and ritonavir 100 mg every 12 h for 14 days (control group). The former group showed negative nasopharyngeal swab in an appreciably short time duration of seven days from start of study in comparison to the control group which took 12 days. [81,82] Lopinavir, a protease inhibitor, inhibits the action of a SARS-CoV-2 enzyme 3-chymotrypsin-like protease ($3CL^{pro}$) which is crucial for viral replication. Ritonavir and Lopinavir are generally administered together to enhance the half-life ($T_{1/2}$) by inhibiting metabolising enzyme cytochrome P450 3A. [83]

9.5.3.3 Ribavirin and Favipiravir

In addition to already discussed therapeutic moieties used in SARS-CoV infection, some newer therapeutic agents have showed promising results against SARS-CoV-2. [84] Few of them have been mentioned below:

Aerosolized Ribavirin (Virazole®, Bausch Health Companies Inc., Canada), an antiviral agent, which acts by inhibition of viral replication is commonly used in infants and adolescents for severe lower respiratory tract infections. [85,86] Ribavirin is a guanosine analogue that prevents RNA and DNA viruses from replicating. When administered to people in combination with other antiviral drugs, it disrupts viral replication in MERS-CoV. *In vitro* studies have shown that it has antiviral effect against the respiratory syncytial virus, influenza and parainfluenza viruses, and US-FDA has recommended it for patients with respiratory syncytial virus, despite the fact that it has no antiviral activity against SARS-CoV. [87, 88] Ribavirin has been generally used in combination with other antivirals for treatment; however, it cannot be used as a stand-alone drug to treat COVID-19.

Similarly, the effectiveness of Favipiravir (Avigan®, FUJIFILM Toyama Chemical Co., Ltd., Tokyo, Japan), a viral RNA-dependent RNA polymerase inhibitor, was studied in a phase 2 clinical trial in Japan. The primary endpoints, alleviation in symptoms such as oxygen saturation, body temperature, could be reached supporting its use as oral drug in SARS-CoV-2 patients. It has also been used in China for treatment of influenza and can stop replication of other RNA virus, such as arenavirus, filoviruses and bunyaviruses. Favipiravir is converted into phosphoribosylated form in cells, which eliminates viral polymerase of RNA. It also inhibits SARS-CoV-2 replication *in vitro*. [88] As compared to the control group treated with umifenovir (or Arbidol), favipiravir exhibited faster clearance of cough and fever but identical incidence of respiratory arrest in an open-label randomised trial. The favipiravir arm exhibited improved outcomes with respect to disease recurrence and viral clearance than the control arm in an open-label non-randomised control study of 80 patients of COVID-19. [89] Various clinical trials of hydroxychloroquine/chloroquine, remdesivir and lopinavir/ritonavir for the treatment of COVID-19 have been listed in Table 9.2.

TABLE 9.2
List of Ongoing Clinical Trials in Various Countries

S.No	Clinical trials	Drug evaluated	Clinical study phase	Status	Country	Ref
1.	NCT04334967	Hydroxychloroquine in COVID-19 diagnosed patients and compared to standard care	Phase 4	Suspended	United States	[173]
2.	NCT04316377	Hydroxychloroquine in adult COVID-19 patients	Phase 4	Active, not recruiting	Norway	[174]
3.	NCT04332094	Hydroxychloroquine vs. azithromycin vs. tocilizumab in COVID-19 hospitalised patients	Phase 2	Recruiting	Spain	[175]
4.	NCT04307693	Hydroxychloroquine or lopinavir/ritonavir in mild COVID-19 patients	Phase 2	Terminated	South Korea	[176]
5.	NCT04321278	Hydroxychloroquine and azithromycin studied in patient with severe pneumonia	Phase 3	Completed	Brazil	[177]
6.	NCT04323527	Various regimens of chloroquine in SARS-CoV-2 patients	Phase 2	Completed	Brazil	[178]
7.	NCT04292899	Phase 3 study: evaluating remdesivir regimens in severe COVID-19 patients	Phase 3	Completed	worldwide sites, including the United States	[179]

(*Continued*)

TABLE 9.2 (*Continued*)
List of Ongoing Clinical Trials in Various Countries

S.No	Clinical trials	Drug evaluated	Clinical study phase	Status	Country	Ref
8.	NCT04292730	Phase 3 study: of different remdesivir regimens moderate COVID-19 patients	Phase 3	Completed	worldwide sites, including the United States	[180]
9.	NCT04280705	Remdesivir vs. placebo in hospitalised patients	Phase 3	Completed	United States, Korea, Japan and Singapore	[181]
10.	DisCoVeRy (NCT04315948)	Lopinavir/ritonavir, remdesivir, hydroxychloroquine and combination of interferon beta-1a in hospitalised COVID-19 patients	Phase 3	Active, not recruiting	France	[182]
11.	NCT04252664	Remdesivir vs. placebo in hospitalised mild and moderate COVID-19 patients	Phase 3	Suspended	China	[183]
12.	NCT04257656	Remdesivir and placebo in hospitalised severe COVID-19 patients	Phase 3	Terminated	China	[184]
13.	NCT04276688	Lopinavir/ritonavir, ribavirin and interferon beta 1b in hospitalised patients	Phase 2	Completed	Hong Kong	[185]
14.	ELACOI (NCT04252885)	Lopinavir/ritonavir vs. arbidol in COVID-19 patients	Phase 4	Completed	China	[186]

9.5.3.4 Arbidol

Arbidol is a chemically synthesised antiviral medication used to treat seasonal influenza. Arbidol has been found to be effective against a variety of virus families. *In vitro*, arbidol exhibits antiviral activity against SARS-CoV during the initial viral replication period. It also stimulates the release of interferons. *In vitro* studies have revealed that arbidol is also effective against COVID-19. [90] Arbidol works by blocking hemagglutinins, a virus protein that binds to sialic acid receptor on human cells and allows the virus to enter the cell via endocytosis, thereby preventing the infection. [91] During an influenza epidemic, in a randomised controlled trial oral arbidol (200 mg/d) was given to workers for 10 to 18 days and strong preventive effects were reported upon administration of arbidol. Similarly, oral arbidol has been used to combat viral infections in people with asthma and chronic obstructive pulmonary disease (COPD) and has also been found to have positive therapeutic outcomes in COVID-19 patients. [92]

9.5.4 Other Category Drugs

9.5.4.1 Disulfiram

Disulfiram is a medicine authorised by the United States Food and Drug Administration (USFDA) for its use in alcohol aversion therapy. It is widely recognised to have an irreversible inhibitory effect on hepatic aldehyde dehydrogenase. According to studies, disulfiram can inhibit other enzymes like urease, kinase and methyltransferase by reacting with cysteine residues, indicating broad-spectrum properties. According to an *in vitro* study, disulfiram inhibits PLpro papin like protease in SARS-CoV and MERS-CoV. [93] In a mouse model of sepsis, Hu et al. discovered that disulfiram has a strong anti-inflammatory effect and reduces mortality, implying that it could be used to treat severe COVID-19 disease. [94] This compound is still in preclinical testing, but it has potential to be repurposed as an anti-coronavirus drug, and there is a need to study its clinical efficacy in treatment of COVID-19. [95]

9.5.4.2 Ivermectin

Ivermectin, an anti-parasitic drug approved by the FDA, blocks SARS-CoV-2 replication *in vitro* by preventing entry of viral proteins into the host cell nucleus. [96] The conditions in which virus replicates and infects the cells differs under the *in vivo* and *in vitro* conditions, so it cannot be defined how ivermectin is useful to patients at this point. Likewise, any pharmacokinetic variations and any unidentified drug interactions that might occur under these conditions are yet to be recognised and observed. Ivermectin, however, may have an advantage over other pharmacotherapeutic options for the treatment of COVID-19 infection. [97] Because of its antiviral activity, ivermectin may play a key role in a number of important biological processes, making it a promising candidate for the management of a variety of viruses, including COVID-19. Clinical trials are required to assess the effect of ivermectin on COVID-19, and this necessitates further research for potential human benefits in current and future pandemics. [98] Caly et al. recently published a report on antiviral activity of ivermectin against COVID-19-virus. [63] It was observed that one dose of ivermectin reduced SARS-CoV-2 replication in Vero/hSLAM cells by 5000-fold. This exploration has attracted the attention of doctors, experts and authorities in public health around the world. Moreover, these outcomes should be cautiously translated.

First, the drug was evaluated only *in vitro* on a single line of monkey kidney cells expressing human signalling lymphocytic activation molecule (SLAM), and it also binds to measles virus receptor CDw150. [99] Ivermectin is yet to be tested in any respiratory cell lines, which are essential for the human SARS-CoV-2 virus. [100] Interestingly, the drug concentration employed in the study (5M) to disrupt SARS-CoV-2 is approximately 35 times greater than the FDA-approved level for treatment of parasitic infections, raising questions regarding its efficacy in human beings using the FDA-approved dose in clinical trials. [101]

9.5.4.3 Azithromycin

Azithromycin belongs to macrolide class of antibiotics that is effective against both gram-positive and gram-negative bacteria. Furthermore, it has shown to have anti-inflammatory and immunoregulatory properties and modulates both the innate immunity and adaptive immunity responses showing effectiveness in the treatment of chronic inflammatory disorders like diffuse bronchiolitis, post-transplant bronchiolitis, non-eosinophilic asthma and rosacea. Azithromycin has also shown better treatment outcome in other viral pneumonias, like influenza or rhinovirus, as well as in patients with acute lung injury admitted to the Intensive Care Unit (ICU). [102] *In vitro* and *in vivo* experiments have shown that macrolides reduce inflammation and simulate the immune system by downregulating cell surface adhesion molecules, reducing the production of pro-inflammatory cytokines, stimulating alveolar macrophage phagocytosis and inhibiting neutrophil activation and mobilisation; however, the mechanism is still unknown. [103] Macrolides, being anti-inflammatory and immunomodulatory, can be used as promising therapeutic agents for the treatment of COVID-19. [104] Azithromycin interferes with binding of SARS-CoV-2 to respiratory cells, as the SARS-CoV-2 spike protein contains a site that binds to the ganglioside and can disrupt SARS-CoV-2 by attaching at this site. This leads to inhibition of virus spike protein from reaching gangliosides on the plasma membrane of host cells, which are important in the pathogenesis of S—S-CoV-2. Furthermore, azithromycin disrupts the interaction between the spike protein and CD147, as well as CD147 expression. [105,106] A retrospective study of 349 MERS patients, 97 of whom received azithromycin, found that macrolide therapy had no effect on 90-day mortality or MERS-CoV RNA clearance. In a recent open-label, non-randomised clinical investigation of SARS-CoV, 20 patients were given hydroxychloroquine, and six of them were given azithromycin to avoid bacterial superinfections. When hydroxychloroquine and azithromycin were given together, SARS-CoV-2 clearance was higher than when given alone. However, in a subsequent study on 11 patients, the same administration protocol produced mixed results, with only one patient reporting a prolonged QT interval and no evidence of increased antiviral activity or clinical benefits. A number of other studies compared the combination of azithromycin and hydroxychloroquine to standard care or hydroxychloroquine alone. Most of them failed to demonstrate a clinical benefit, instead emphasised on the risk of having corrected QT (QTc) prolongation. [104]

9.5.4.4 Doxycycline

Doxycycline is efficacious in decreasing inflammatory factors such as IL-6 and TNF-alpha, which are responsible for triggering cytokine storms and likely leading to death in COVID-19 patients. It has few side effects and can be taken orally. [107] Doxycycline treatment has been shown to provide protection against lung injury in experimental studies. The use of doxycycline as a preventive measure in infected mice with the virulent influenza H3N2 virus reduced the incidence of acute lung injury. Independent of their antimicrobial activities, the tetracycline group of antimicrobial agents have been found to be therapeutically useful in inhibition of matrix metalloproteinases (MMP) due to their ability to chelate the catalytic Zn^{2+} ion, which is required for MMP activity. Doxycycline is the most promising MMP inhibitor amongst the tetracycline derivatives, even at sub-antimicrobial doses (25 mg). Since lung immune injury/ARDS is common in patients with severe COVID-19, suppressing MMPs may aid in lung tissue regeneration and repair. [108] Doxycycline affected both SARS-CoV-2 entry as well as virus replication after infection and can be used as a potential COVID-19 drug. Further, testing of doxycycline on animal experiment models as well as in patients with moderate to severe COVID-19 infection will be useful. [109]

9.5.4.5 Corticosteroids

Regardless of evidence of their therapeutic potential, corticosteroids have been routinely given to COVID-19 patients to reduce both lung inflammation and incidence of ARDS. Moreover, a significant delay in viral clearance, and at times, increased risk of secondary infection has been observed, both of which could be serious drawbacks. In patients with COVID-19, corticosteroids should be used with caution due to lack of evidence of benefit and the risk of harm. [89]

9.5.4.5.1 Methylprednisolone

It is a corticosteroid drug, which has powerful anti-inflammatory properties and may be useful for the treatment of pneumonia. However, the evidence of its effectiveness in coronavirus infections is inconclusive and disputed. In addition, the risk factors of secondary infection and delayed viral clearance frequently outweigh the advantages. [110,111]

9.5.4.5.2 Dexamethasone

It's a type of glucocorticoid that is used to cure inflammatory and autoimmune diseases. Preliminary data from a recovery study showed reduction in mortality with dexamethasone by one-third in mechanically ventilated hospitalised patients having severe COVID-19 infection and by one-fifth in patients who required oxygen without mechanical ventilation. In patients without respiratory support, this medicament did not improve survival. The use of dexamethasone has been approved in UK for the treatment of COVID-19. As with other corticoids, dexamethasone weakens the immune system and may thereby increase the potential of

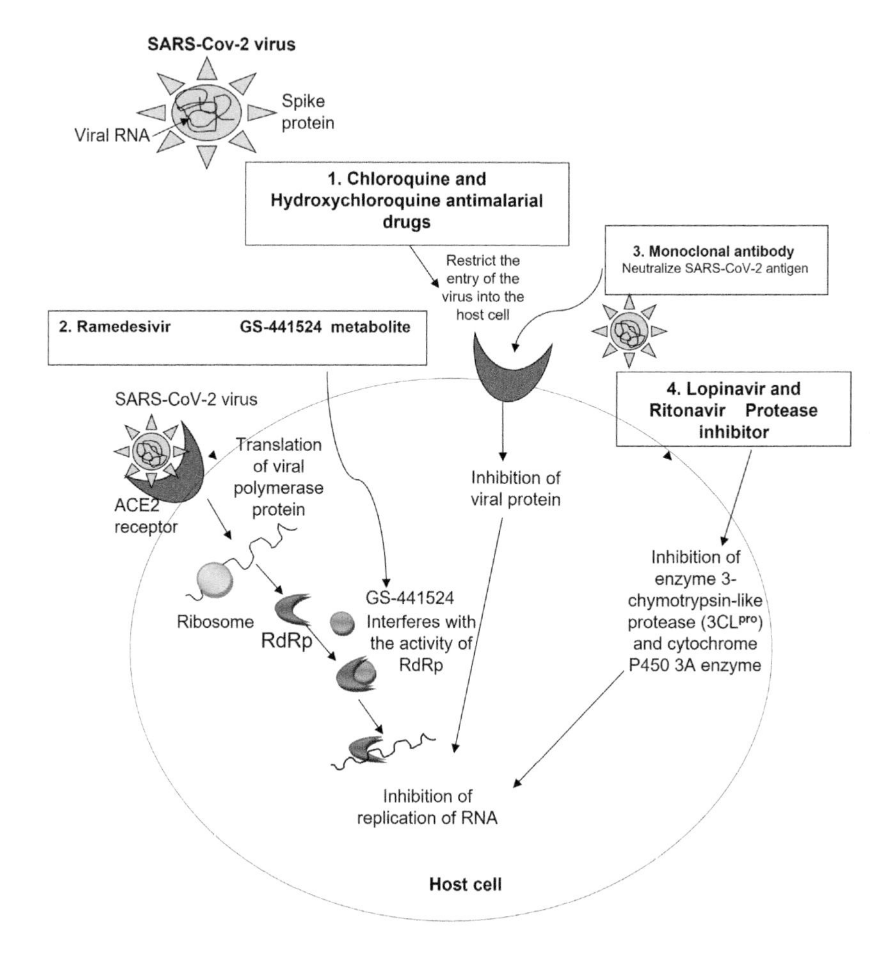

FIGURE 9.3 Diagrammatic representation of the mechanism of action of therapeutic agents against the SARS-CoV-2 virus in the host. [2], [3]

viral infections. However, its immunosuppressive properties could help COVID-19 patients with cytokine release syndrome (CRS). [88]

9.5.5 Nutritional Supplements

A balanced diet is essential to keep the immune system fit and active. Persons with malnutrition have a greater chance of being infected, and nutrient deficiency can modify immune response. Optimal nutrition can improve well-being and may help to reduce the risk and morbidity due to COVID-19. [112] Many nutrients are required for strengthening the immune system to function properly.

9.5.5.1 Vitamin C

Vitamin C, also called ascorbic acid, is essential for a variety of physiological effects in the body. In systemic inflammatory syndrome, T lymphocytes and NK cells play a significant role in the development of immunity against viral infections, limiting generation of reactive oxygen species and remodelling of cytokine network as shown in **Figure 9.4.** Vitamin C improves immunity by enhancing production of interferons, proliferation of lymphocyte and phagocytosis by neutrophil. There is clear evidence of enhancement of immunity upon use of vitamin C, against SARS-CoV-2 virus including COVID-19. [113,114]

9.5.5.2 Vitamin D

Vitamin D is generated by the human body in response to sunlight, seems to play significant functions in adaptive and innate immunity as shown in **Figure 9.5,** along with immune cell differentiation, proliferation and maturation. Due to these functions, vitamin D has been tested for new SARS-CoV-2 infections in clinical trials as an immune modulator. [115,116]

9.5.5.3 Folic Acid

Folic acid is also known as vitamin B9, which is required for production of purines, pyrimidines and methionine, which are considered necessary for the formation of RNA, DNA and proteins. Although the human body cannot produce folic acid, it is an important nutrient that must be supplied through food. [117]

9.5.5.4 Probiotics

Probiotics are live microbes that have health benefits for the host when administered in an appropriate amount. They include a variety of bacteria and yeast. Lactobacillus, Bifidobacterium, Leuconostoc, Pediococcus and Enterococcus

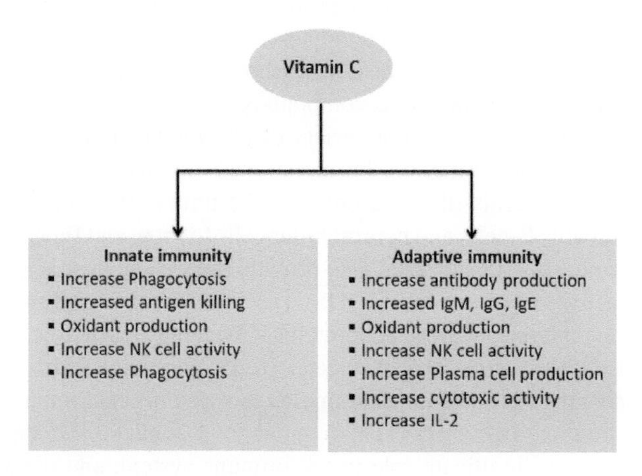

FIGURE 9.4 Role of Vitamin C in innate and adaptive immunity. [4]

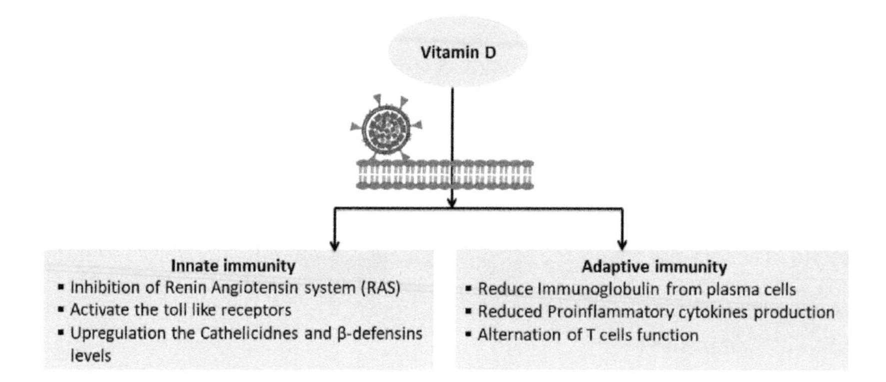

FIGURE 9.5 Role of vitamin D in development of innate and adaptive immune response against coronavirus. [5]

are only a few of the probiotic bacteria. Microflora belonging to the genera Lactobacillus and Bifidobacterium populate the normal gastrointestinal microflora of humans. They're perfectly safe and can be found in a variety of yogurts and other dairy products. [118] Probiotics have been shown to have therapeutic benefits against a number of diseases. Probiotics play a role in regulating the immune response of the host. They also regulate the feature of systemic immune cells, intestinal epithelial cells and the mucosal immune system, as indicated by studies on host immunity. As a result, probiotics have the potential to be therapeutically used in diseases involving immune system, such as viral infections, allergies, eczema and vaccination response. [119]

9.5.5.5 Minerals

In the lack of appropriate treatment, a powerful immunity is one of the most efficient protective mechanisms when it comes to COVID-19 infection.

Mineral supplementation has also been shown to improve immunity to viral infections. Minerals are inorganic compounds that the body needs to function properly. Minerals play a role in a variety of physiological processes, including bone formation, blood formation, hormone synthesis and heartbeat regulation. [120] Numerous minerals like zinc enhance the immunity. Zinc is required for neutrophils, T cells, B-cells and natural killer cells to grow and function normally. The antioxidant properties of Zinc also provide protection against inflammatory effects of reactive oxygen species (ROS). [121] Selenium is a potent nutritional antioxidant that interacts with selenoproteins. Given the critical role of selenoproteins in controlling reactive oxygen species (ROS) and redox status in almost all tissues, it's no surprise that the selenium diet has a significant impact on inflammation and immune response. [122] A number of studies have suggested that phosphorus plays a significant role in the immune system, and dietary changes in phosphorus levels could affect immune cell function and migration. Calcium, too, is a normal body component that is essential for immune function. [123] Iron improves phagocytosis by regulating cytokine production. Copper promotes neutrophil phagocytosis and increases the production of interleukin-2 (IL-2). [124] Minerals are therefore essential for a strong immune function, and enhancing the body's mineral content could be a helpful means to promote defence against COVID-19 infection.

9.5.6 Use of Disinfectants and Soaps

In favourable atmospheric conditions, SARS-CoV-2 can survive for up to 72 hours on surfaces such as plastic and stainless steel, but it is prone to numerous disinfectants such as sodium hypochlorite, 75% ethanol, hydrogen peroxide, chloroform, diethyl ether and others. Soaps are similarly efficient since they rapidly dissolve the cell membrane of the virus. UV light can also deactivate SARS-CoV-2, besides application of heat at 60°C for 30 minutes. [88]

9.5.7 Vaccines Against SARS-CoV-2

Scientists of National Institute of Allergy and Infectious Diseases (NIAID) have been constantly trying to develop vaccines to prevent entry of virus, block viral replication, delay immune system responses and also build immunity against SARS-CoV-2 infection. The vaccine, mRNA-1273 has been co-developed by scientists of NIAID and Moderna, Inc., (Cambridge, Massachusetts) and entered late-stage clinical trials in July 2020. The US FDA issued an Emergency Use Authorization (EUA) for the Moderna COVID-19 Vaccine (mRNA-1273) on December 18, 2020, for administration to patients of 18 years of age and older. [125–127] NIAID has also conducted a phase 1 clinical trial for two therapeutic agents SAB-301 (for treatment of MERS) and combination of two monoclonal antibodies REGN3048 and REGN 3051, both of which have been found to be safe and effective. [128–130] INOVIO has developed a DNA vaccine INO-4800, which is currently in phase 1 clinical study, for which $17.2 million funds have been partly funded by Coalition for Epidemic Preparedness Innovations and is currently preparing for phase 3 of clinical trial which has yet to be approved by regulatory bodies. [131] For the prevention of SARS-CoV-2, AstraZeneca in collaboration with the University of Oxford, had announced global development and distribution of its potential recombinant adenovirus vaccine. [132] On May 21, 2021, the Japanese Ministry of Health, Labour and Welfare had granted approval for the emergency use of AstraZeneca's COVID-19 vaccine, that is, Vaxzevria (ChAdOx1-S) which was formerly referred to as AZD1222 in Japan for the candidates 18 years or more as a preventive measure to curb the spread of the COVID-19 disease. [133] Japan's Pharmaceuticals and Medical Devices Agency recommends two 0.5 mL doses of Vaxzevria administered intramuscularly at an interval of four to twelve weeks (an interval of more than eight weeks is preferred to accentuate the efficacy). The vaccine can be stored for at least six months at 2–8° C (36–46° F) and can also be transported and handled at the same storage conditions. The vaccine employs a replication deficient chimpanzee viral vector based on a weakened version of adenovirus (common cold virus) incorporating the genetic material for the spike protein of the SARS-CoV-2 virus, which expresses itself after immunisation initiating an immune response in vaccinated person. [133,134] The data from PHE (Public Health England) demonstrated that AstraZeneca COVID-19 vaccine also offers protection against the Delta variant (B.1.617.2) which was formerly known as the 'Indian' variant with 92% efficacy and the Alpha variant (B.1.1.7) which was known as 'Kent' variant with 86% efficacy. [135] ChAdOx1 vaccine manufactured by Serum Institute of India Pvt Ltd as COVISHIELD is given in two doses at an interval of 12–16 weeks. [136] Other pharmaceutical companies, like Johnson & Johnson, Sanofi, Pfizer and BioNTech, also took part effectively in the development of a safe and effective vaccine. [137–140]

9.5.7.1 Janssen Pharmaceutical Company of Johnson & Johnson

An emergency use authorization (EUA) issued by US FDA on February 27, 2021, authorised the use of COVID-19 vaccine developed by the Janssen Pharmaceutical Company of Johnson & Johnson in candidates 18 years of age and older as a preventive measure for the COVID-19. The single dose vaccine has been reported to be stable for a period of two years at the storage condition of -4°F (-20°C) and maximum for a period of three months at 36–46°F (2–8°C). It has been notified that the vaccine should not be refrozen if allowed to be distributed at temperatures of 36–46°F (2–8°C). The Janssen COVID-19 vaccine grounds on the AdVac® vaccine platform that was also deployed for the development and manufacture of Janssen's Ebola vaccine regimen approved by the European Commission. [141]

9.5.7.2 Bharat Biotech

In collaboration with the ICMR-NIV (Indian Council of Medical Research—National Institute of Virology) Bharat Biotech developed COVAXIN® (an inactivated vaccine) based on Whole Virion Inactivated Vero Cell derived technology. The vaccine is meant to be administered twice, with 28 days interval between the consecutive doses and should be stored at conditions of 2–8° C without requirement of sub-zero storage. COVAXIN® contains immune potentiators or vaccine adjuvants that are meant to accentuate the immunogenicity of the vaccine. In July 2020 DCGI (Drugs Controller General of India) approval was received for the vaccine for phase 1 and phase 2 human clinical trials. [142–145] On June 22, 2021, the Subject Expert Committee (SEC) under DCGI reviewed the data of phase 3 clinical trial submitted by Bharat Biotech in which efficacy of 77.8% was observed for the COVAXIN®. [146] The detailed information on COVAXIN® can be found at these references. [142,144,147–150]

9.5.7.3 Gamaleya (Sputnik V)

Russian Ministry of Health, the Gamaleya National Research Center of Epidemiology and Microbiology and the Russian Direct Investment Fund Sputnik

Light (a single dose COVID-19 vaccine of Sputnik V) authorised its use in Russia. [151] This vaccine has been named after the first Soviet Space Satellite that is Sputnik-1 that was launched in the year 1957, instigating a so-called Sputnik moment around the world. [152] Sputnik V is a recombinant human adenovirus vaccine and contains two different adenovirus vectors which makes this vaccine unique in the series of COVID-19 vaccines developed so far around the world, imparting lasting immunity. The first component of Sputnik V, that is, Sputnik Light is a recombinant human adenovirus serotype number 26 referred to as rAd26 and the second component is a recombinant human adenovirus serotype number 5 (rAd5) which is stable at the storage conditions of 2–8° C. The data obtained from Russia's mass vaccination programme that was held from December 5, 2020 to April 15, 2021 revealed an efficacy of 79.4% for Sputnik Light with effectiveness against all new strains of coronavirus as proven by the laboratory tests of Gamaleya Center. Based on the analysis of data obtained from 19,866 volunteers who received both the doses of the Sputnik V or placebo at the final control point with 78 confirmed SARS-CoV2 cases, Sputnik V demonstrated an efficacy level of 91.6 %. [152–154]

9.5.7.4 Novavax

Based on Novavax recombinant nanoparticle technology and its patented saponin based Matrix-M™ adjuvant (that tends to accentuate the immune response and instigate the production of high levels of neutralising antibodies by inciting the entry of APCs (antigen-presenting cells) towards the site of injection and thus provoking the presentation of antigen into local lymph nodes), Novavax has developed a protein-based vaccine NVX-CoV2373, which is genetically engineered from sequence of the first strain of SARS-CoV2 encoding coronavirus spike protein. [155] Immune response induced by NVX-CoV2373 halts the binding of spike protein to the cellular receptors as observed in the preclinical studies. Data evaluated from the pivotal phase 3 trials (one in UK and other one in the US and Mexico referred to as PREVENT-19 trial) of NVX-CoV2373 showed an efficacy of 96.4 % against the original virus strain and 86.3 % against the Alpha variant, that is, B.1.1.7, and an overall efficacy of 89.7 %. NVX-CoV2373 is stable and can be stored at 2–8° C. [156–158]

9.5.7.5 Sinopharm

The vaccine developed by the Sinopharm/Beijing Bio-Institute of Biological Products Co Ltd, which is a subsidiary of China National Biotec Group (CNBG) received Emergency Use Listing (EUL) by the WHO on May 7, 2021. [159] The Sinopharm vaccine is an inactivated vaccine available in two doses of 0.5 mL, each administered intramuscularly within an interval of three to four weeks as recommended by WHO. An efficacy level of 79 % for the two-dose regimen was exhibited by the Sinopharm vaccine based on a multi-country phase 3 trial when administered at an interval of 21 days between the consecutive doses. [160]

9.5.7.6 Biontech/Pfizer

The first Emergency Use Authorization (EUA) was issued by the US FDA dated December 11, 2020, for the COVID-19 vaccine in the candidates age 16 years and older developed by the Biontech/Pfizer. [161,162] Further, US FDA expanded the EUA for Biontech/Pfizer COVID-19 vaccine for candidates of age 12 to 15 on May 10, 2021. [163] The EUA allows the Biontech/Pfizer COVID-19 vaccine to be distributed in the United States.

9.5.7.6.1 Herbal and Ayurvedic Treatment

Amongst various approaches, attention has also been directed towards herbal and ayurvedic formulations. *Artemisia Annua*, also known as sweet wormwood, is used for the treatment of malaria, and the scientists at Germany's Max Planck Institute of Colloids and Interfaces in Potsdam in collaboration with the US Company ArtemiLife have explored the use of *Artemisia* plant in coronavirus infection. In a study conducted in April 2020, artemisinin, active constituent isolated from *Artemisia*, was found to be more effective than hydroxychloroquine. [164] An international collaboration between Department of Biotechnology (DBT) and the National Institute of Advanced Industrial Science and Technology (AIST), Japan at Indian Institute of Technology, Delhi (IIT Delhi) proposed that bioactives of Ashwagandha (*Withania somnifera*) and propolis can be used in severe acute respiratory syndrome occurring in SARS-CoV-2. Withanone and caffic acid phenethyl ester (CAPE), active ingredients from Ashwagandha and New Zealand propolis respectively target main protease (Mpro) enzyme that restrains the replication of SARS-CoV-2 and also blocks TMPRSS2. [165] *ArtemiC*, a micellar formulation comprising curcumin and artemisinin jointly developed by MGC Pharmaceuticals and Switzerland firm Micelle Technology, was studied in a clinical trial as oral spray for its anti-viral, anti-inflammatory and anti-oxidant properties in SARS-CoV-2 patients. Although the clinical trial has been completed, results have not been published. [166] Ministry of AYUSH, India has suggested some self-care measures for enhancing immunity for protection against the infection, which includes intake of golden milk prepared by adding half teaspoon turmeric powder in 150 ml milk, herbal tea/decoction (Kadha) made from basil, cinnamon, black pepper, dry ginger and raisin with or without honey or jaggery, once or twice a day. [167] Use of nutraceuticals, dietary supplements and traditional medicines is also being explored for combatting SARS-CoV-2. In a randomised, multicenter and doubled armed phase 3 clinical trial, in order to potentially reduce rates of hospitalisation and improve clinical recovery time and symptoms in non-hospitalised coronavirus positive patients, short-term administration of a dietary supplements containing resistant starch was studied. Though the trial is over, the results of the trial are still to be published. [168] Similarly, to test the efficacy of natural honey in infected patients, a randomised, multicenter, controlled trial was conducted in comparison to standard care, but the results are awaited. [169] According to the guidelines of the National Institute for Health and Care Excellence (NICE) and the Public Health England (PHE), natural honey has the potential to treat acute cough caused

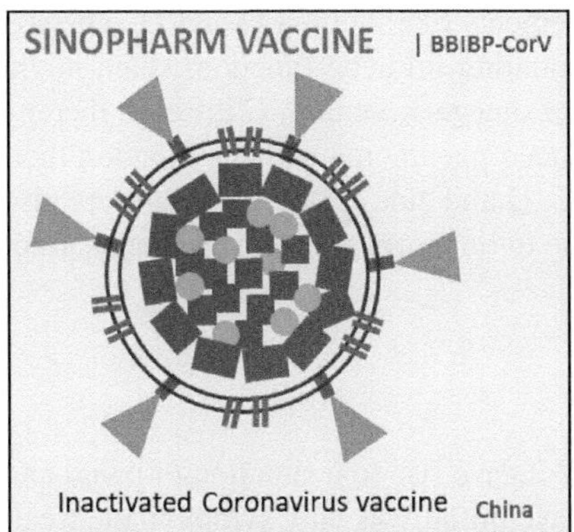

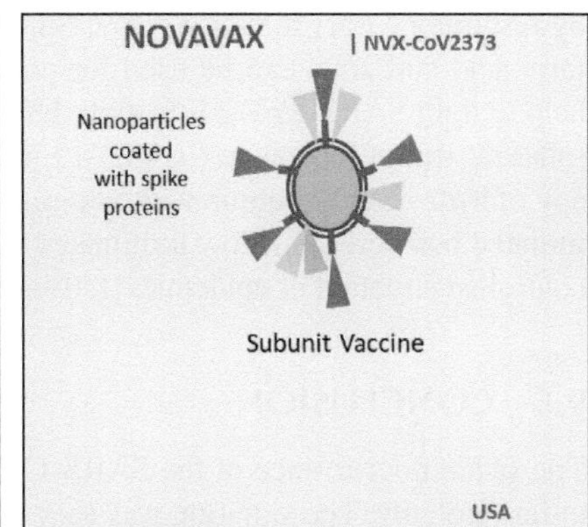

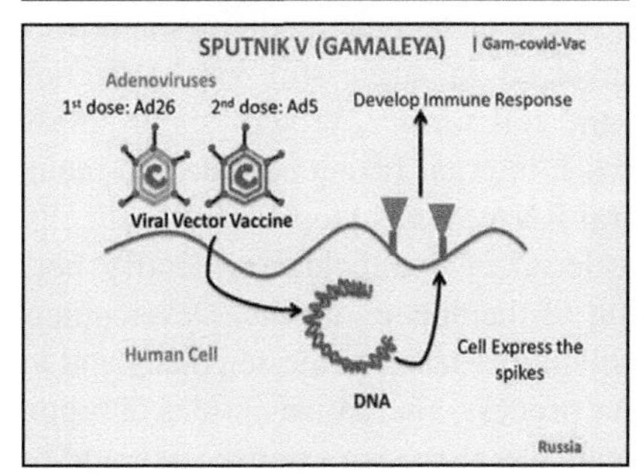

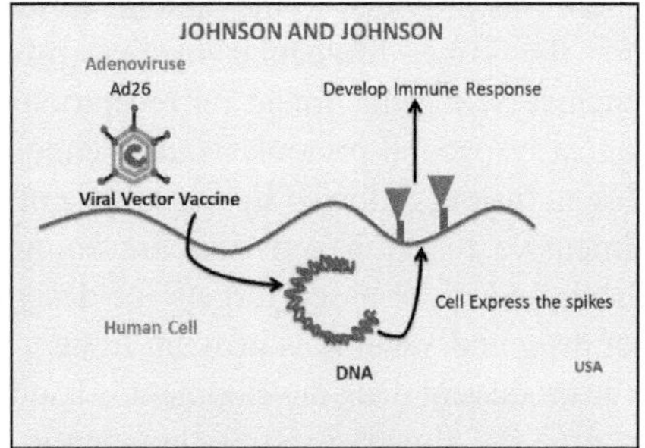

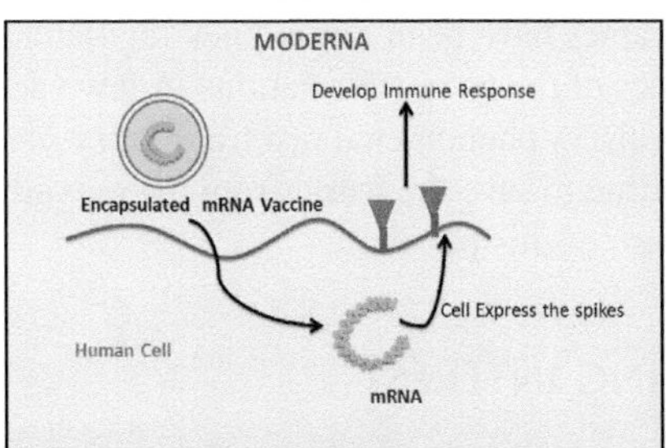

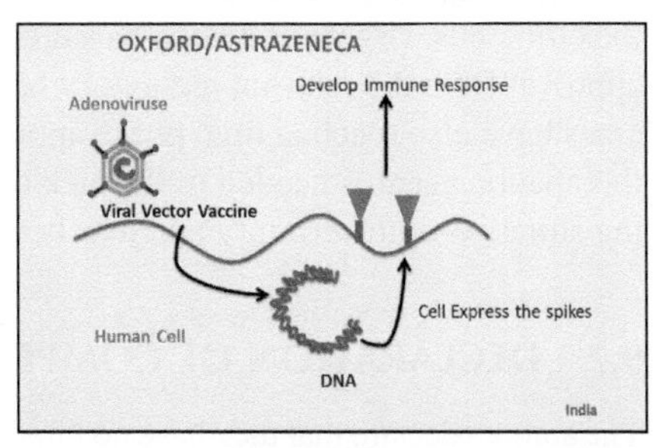

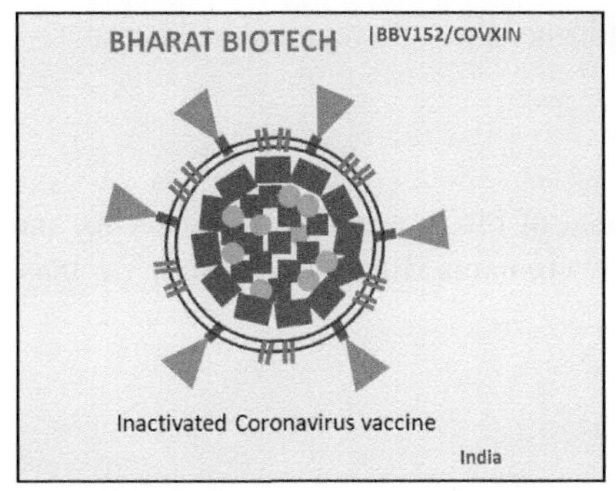

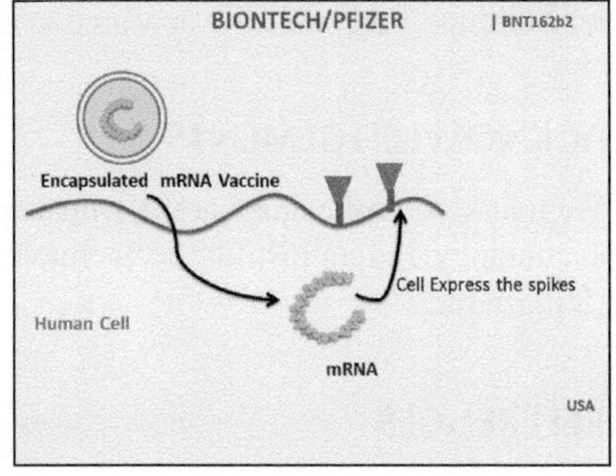

FIGURE 9.6 Pictorial representation of mechanism of action of various vaccines developed against COVID-19.

by respiratory tract infection. [169] Supplements like vitamin C and D, omega-3 fatty acid and zinc, can be used for potentiating immune function which might help to fight SARS-CoV-2 infection. Use of omega-3, vitamin C infusion therapy (phase 2 study) for severe COVID-19 induced pneumonia and combination therapy of hydroxychloroquine with vitamin C, D and zinc (phase 2 study) were also initiated but were somehow terminated due to improper recruitment as a result of controlled situation of epidemic. [170–172]

9.6 CONCLUSION

The sudden occurrence of the SARS-CoV-2 gave rise to a situation of panic and unrest globally. The situation was very grave initially as only symptomatic treatment was provided to the patients in the absence of specific medicines approved for the same. Molecular mechanisms involved in successful viral infection include viral attachment on receptors on the cell surface, proteolysis or subsequent lysosomal proteolysis after endocytosis, internal fusion peptide and membrane fusion, followed by the release of viral RNA to the cytoplasm of host cells. Intensive research activities are being undertaken worldwide to identify both host- and virus-based targets for designing of therapeutic agents. Development of drug and vaccine is proving to be a challenging task for the scientists and as the process of drug development is a tedious process, alternative options of repurposing the already existing therapeutic agents has also been extensively opted by the clinicians. Urgent and need-based therapies have been given quick regulatory approval from the relevant regulatory bodies of the respective nations. A few vaccines have also reached final phase approvals. A phenomenal fast track entry of a therapeutic agent is needed in the nick of time to save the world from the prevailing situation of uncertainty and fear because of this pandemic.

9.7 DECLARATION OF COMPETING INTEREST

The authors declare that they have no known competing financial interests or personal relationships that could have appeared to influence the work reported in this paper.

ACKNOWLEDGEMENTS

We thank our colleagues at the Department of Pharmaceutical Engineering and Technology, Indian Institute of Technology (Banaras Hindu University), Varanasi (UP), India.

REFERENCES

1. Song, Z., Xu, Y., Bao, L., Zhang, L., Yu, P., Qu, Y., Zhu, H., Zhao, W., Han, Y., Qin, C. 2019. From SARS to MERS, thrusting coronaviruses into the spotlight. Viruses. MDPI AG. 11(1), 59–87.

2. Li, G., De Clercq, E. 2020. Therapeutic options for the 2019 novel coronavirus. Nat Rev Drug Discov. NLM (Medline). 19, 149–150.
3. Wang, M., Cao, R., Zhang, L., Yang, X., Liu, J., Xu, M., Shi, Z., Hu, Z., Zhong, W., Xiao, G. 2020. Remdesivir and chloroquine effectively inhibit the recently emerged novel coronavirus (2019-nCoV) in vitro. Cell Res. 30, 269–271.
4. Zheng, J. 2020. SARS-coV-2: An emerging coronavirus that causes a global threat. Int J Biol Sci. 16(10), 1678–1685.
5. Gorbalenya, A.E., Baker, S.C., Baric, R.S., de Groot, R.J., Drosten, C., Gulyaeva, A.A., Haagmans, B.L., Lauber, C., Leontovich, A.M., Neuman, B.W., Penzar, D., Perlman, S., Poon, L.L.M., Samborskiy, D.V., Sidorov, I.A., Sola, I., Ziebuhr, J. 2020. The species Severe acute respiratory syndrome-related coronavirus: Classifying 2019-nCoV and naming it SARS-CoV-2. Nat Microbiol Nat Res. 5, 536–544.
6. Venkat, K.G., Jeyanthi, V., Ramakrishnan, S. 2020. A short review on antibody therapy for COVID-19. New Microbes New Infect. 35, 100682.
7. Perlman, S., Netland, J. 2009. Coronaviruses post-SARS: Update on replication and pathogenesis. Nat Rev Microbiol. 7, 439–450.
8. Ye, Q., Wang, B., Mao, J. 2020. The pathogenesis and treatment of the 'cytokine storm' in COVID-19. J Infect. 80, 607–613.
9. Peiris, J.S.M., Lai, S.T., Poon, L.L.M., Guan, Y., Yam, L.Y.C., Lim, W., Nicholls, J., Yee, W.K.S., Yan, W.W., Cheung, M.T., Cheng, V.C.C., Chan, K.H., Tsang, D.N.C., Yung, R.W.H., Ng, T.K., Yuen, K.Y. 2003. Coronavirus as a possible cause of severe acute respiratory syndrome. Lancet. 361(9366), 1319–1325.
10. Kuiken, T., Fouchier, R.A.M., Schutten, M., Rimmelzwaan, G.F., Van, A.G., Van Riel, D., Laman, J.D., De Jong, T., Van, D.G., Lim, W., Ling, A.E., Chan, P.K.S., Tam, J.S., Zambon, M.C., Gopal, R., Drosten, C., Van, D.W.S., Escriou, N., Manuguerra, J.C., Stöhr, K., Peiris, J.S.M., Osterhaus, A.D.M.E. 2003. Newly discovered coronavirus as the primary cause of severe acute respiratory syndrome. Lancet. 362(9380), 263–270.
11. Zaki, A.M., Van, B.S., Bestebroer, T.M., Osterhaus, A.D.M.E., Fouchier, R.A.M. 2012. Isolation of a novel coronavirus from a man with pneumonia in Saudi Arabia. N Engl J Med. 367(19), 1814–1820.
12. Cai, Q., Yang, M., Liu, D., Chen, J., Shu, D., Xia, J., Liao, X., Gu, Y., Cai, Q., Yang, Y., Shen., C., Li, X., Peng, L., Huang, D., Zhang, J., Zhang, S., Wang, F., Liu, J., Chen, L., Chen, S., Wang, Z., Zhang, Z., Cao, R., Zhong, W., Liu, Y., Liu, L. 2020. Experimental treatment with favipiravir for COVID-19: An open-label control study. Engineering. 6(10), 1192–1198.
13. Acter, T., Uddin, N., Das, J., Akhter, A., Choudhury, T.R., Kim, S. 2020. Evolution of severe acute respiratory syndrome coronavirus 2 (SARS-CoV-2) as coronavirus disease 2019 (COVID-19) pandemic: A global health emergency. Sci Total Environ. 730, 138996.
14. Masters, P.S. 2006. The Molecular Biology of Coronaviruses. Adv Virus Res. 65, 193–292.
15. Belouzard, S., Millet, J.K., Licitra, B.N., Whittaker, G.R. 2012. Mechanisms of coronavirus cell entry mediated by the viral spike protein. Viruses. 4, 1011–33.
16. Weiss, S.R., Navas-Martin, S. 2005. Coronavirus pathogenesis and the emerging pathogen severe acute respiratory syndrome coronavirus. Microbiol Mol Biol Rev. 69(4), 635–664.
17. Desforges, M., Desjardins, J., Zhang, C., Talbot, P.J. 2013. The acetyl-esterase activity of the hemagglutinin-esterase protein of human coronavirus OC43 strongly enhances the production of infectious virus. J Virol. 87(6), 3097–3107.

18. Zeng, Q., Langereis, M.A., van Vliet, A.L.W., Huizinga, E.G., de Groot, R.J. 2008. Structure of coronavirus hemagglutinin-esterase offers insight into corona and influenza virus evolution. Proc Natl Acad Sci U S A. 105(26), 9065–9069.
19. Ye, Q., Wang, B., Mao, J. 2020. The pathogenesis and treatment of the 'cytokine storm' in COVID-19. J Infect. W.B. Saunders Ltd. 80, 607–613.
20. Zhang, W., Zhao, Y., Zhang, F., Wang, Q., Li, T., Liu, Z., Wang, J., Qin, Y., Zhang, X., Yan, X., Zeng, X., Zhang, S. 2020. The use of anti-inflammatory drugs in the treatment of people with severe coronavirus disease 2019 (COVID-19): The experience of clinical immunologists from China. Clin Immunol. Academic Press Inc. 214, 108393.
21. Huang, C., Wang, Y., Li, X., Ren, L., Zhao, J., Hu, Y., Zhang, L., Fan, G., Xu, J., Gu, X., Cheng, Z., Yu, T., Xia, J., Wei, Y., Wu, W., Xie, X., Yin, W., Li, H., Liu, M., Xiao, Y., Gao, H., Guo, L., Xie, J., Wang, G., Jiang, R., Gao, Z., Jin, Q., Wang, J., Cao, B. 2019. Clinical features of patients infected with 2019 novel coronavirus in Wuhan, China. Lancet. 395, 497–506.
22. Hoffmann, M., Kleine-Weber, H., Schroeder, S., Krüger, N., Herrler, T., Erichsen, S., Schiergens, T.S., Herrler, G., Wu, N.H., Nitsche, A., Müller, M.A., Drosten, C., Pöhlmann, S. 2020. SARS-CoV-2 cell entry depends on ACE2 and TMPRSS2 and is blocked by a clinically proven protease inhibitor. Cell. 181(2), 271–280.
23. Chen, Y., Liu, Q., Guo, D. 2020. Emerging coronaviruses: Genome structure, replication, and pathogenesis. J Med Virol. John Wiley and Sons Inc. 92, 418–423.
24. Fehr, A.R., Perlman, S. 2015. Coronaviruses: An overview of their replication and pathogenesis. In: Coronaviruses: Methods and Protocols. Springer, New York. 1282, 1–23.
25. Fung, T.S., Liu, D.X. 2014. Coronavirus infection, ER stress, apoptosis and innate immunity. Front Microbiol. Frontiers Research Foundation. 5, 1–13.
26. Marano, G., Vaglio, S., Pupella, S., Facco, G., Catalano, L., Liumbruno, G.M., Grazzini, G. 2016. Convalescent plasma: New evidence for an old therapeutic tool? Blood Transfusion. SIMTI Servizi Sri. 14, 152–157.
27. Rojas, M., Rodríguez, Y., Monsalve, D.M., Acosta-Ampudia, Y., Camacho, B., Gallo, J.E., Rojas-Villarraga, A., Ramírez-Santana, C., Díaz-Coronado, J.C., Manrique, R., Mantilla, R.D., Shoenfeld, Y., Anaya, J.M. 2020. Convalescent plasma in Covid-19: Possible mechanisms of action. Autoimmun Rev. Elsevier B.V. 19, 102554.
28. Von, B.E., Kitasato, S. 1890. The mechanism of immunity in animals to diphtheria and tetanus. Deutscbe Med Wocbens. 16, 1113–1114.
29. Hui, D.S.C., Lee, N. 2013. Adjunctive therapies and immunomodulating agents for severe influenza. Influenza Other Respir Viruses. 7, 52–59.
30. Leider, J.P., Brunker, P.A.R., Ness, P.M. 2010. Convalescent transfusion for pandemic influenza: Preparing blood banks for a new plasma product? Transfusion. 50, 1384–1398.
31. Zingher, A., Mortimer, P. 2005. Convalescent whole blood, plasma and serum in the prophylaxis of measles. Rev Med Virol. 15(6), 1180–1187.
32. Nour, B., Green, M., Michaels, M., Reyes, J., Tzakis, A., Gartner, J.C., McLoughlin, L., Starzl, T.E. 1993. Parvovirus B19 infection in pediatric transplant patients. Transplantation. 56(4), 835–838.
33. Hemming, V.G. 2001. Use of intravenous immunoglobulins for prophylaxis or treatment of infectious diseases. Clin Diagn Lab Immunol. American Society for Microbiology (ASM). 8, 859–863.
34. Brown, B.L., McCullough, J. 2020. Treatment for emerging viruses: Convalescent plasma and COVID-19. Transfus Apher Sci. Elsevier Ltd. 59(3), 102790.

35. Cheng, Y., Wong, R., Soo, Y.O.Y., Wong, W.S., Lee, C.K., Ng, M.H.L., Chan, P., Wong, K.C., Leung, C.B., Cheng, G. 2005. Use of convalescent plasma therapy in SARS patients in Hong Kong. Eur J Clin Microbiol Infect Dis. 24(1), 44–46.
36. Kong, L., Zhou, B. 2006. Successful treatment of avian influenza with convalescent plasma. Hong Kong Med J. 12(6), 489.
37. Abdallat, M.M., Abroug, F., Al Dhahry, S.H.S., Alhajri, M.M., Al-Hakeem, R., Al Hosani, F.I., Al Qasrawi, S.M.A., Al-Romaihi, H.E., Assiri, A., Baillie, J.K.B., Embarek, P.K.B., Salah, A., Blümel, B., Briese, T., Buchholz, U., Cognat, S.B.F., Defang, G.N., De La, Rocque, S., Donatelli, I., Drosten, C., Drury, P.A., Eremin, S.R., Ferguson, N.M., Fontanet, A., Formenty, P.B.H., Fouchier, R.A.M., Gao, C.Q., Garcia, E., Gerber, S.I., Guery, B., Haagmans, B.L., Haddadin, A.J., Hardiman, M.C., Hensley, L.E., Hugonnet, S.A.L., Hui, D.S.C., Isla, N., Karesh, W.B., Koopmans, M., Kuehne, A., Lipkin, W.I., Mafi, A.R., Malik, M., Manuguerra, J.C., Memish, Z., Mounts, A.W., Mumford, E., Opoka, L., Osterhaus, A., John, Oxenford, C., Pang, J., Pebody, R., Peiris, J.S.M.J., Plotkin, B., Poumerol, G., Reusken, C., Rezza, G., Roth, C.E., Shindo, N., Shumate, A.M., Siwula, M., Slim, A., Smallwood, C., Vander, W.S., Van, K.M.D., Zambon, M. 2013. State of knowledge and data gaps of middle east respiratory syndrome coronavirus (MERS-CoV) in humans. PLoS Curr. 5(Outbreaks), 12–16.
38. Zhou, B., Zhong, N., Guan, Y. 2007. Treatment with convalescent plasma for influenza A (H5N1) infection. New Eng J Med. Massachussetts Medical Society. 357, 1450–1451.
39. CFR—code of federal regulations title 21. 2020. www.accessdata.fda.gov/scripts/cdrh/cfdocs/cfcfr/CFRSearch.cfm?CFRPart=630.
40. Duan, K., Liu, B., Li, C., Zhang, H., Yu, T., Qu, J., Zhou, M., Chen, L., Meng, S., Hu, Y., Peng, C., Yuan, M., Huang, J., Wang, Z., Yu, J., Gao, X., Wang, D., Yu, X., Li, L., Zhang, J., Wu, X., Li, B., Xu, Y., Chen, W., Peng, Y., Hu, Y., Lin, L., Liu, X., Huang, S., Zhou, Z., Zhang, L., Wang, Y., Zhang, Z., Deng, K., Xia, Z., Gong, Q., Zhang, W., Zheng, X., Liu, Y., Yang, H., Zhou, D., Yu, D., Hou, J., Shi, Z., Chen, S., Chen, Z., Zhang, X., Yang, X. 2020. Effectiveness of convalescent plasma therapy in severe COVID-19 patients. Proc Natl Acad Sci USA. 117(17), 9490–9496.
41. Tiberghien, P., Lamballerie, X., Morel, P., Gallian, P., Lacombe, K., Yazdanpanah, Y. 2020. Collecting and evaluating convalescent plasma for COVID-19 treatment: Why and how? Vox Sang. 115(6), 488–494.
42. Frame, J.D., Verbrugge, G.P., Gill, R.G., Pinneo, L. 1984. The use of lassa fever convalescent plasma in Nigeria. Trans R Soc Trop Med Hyg. 78(3), 319–324.
43. Shen, C., Wang, Z., Zhao, F., Yang, Y., Li, J., Yuan, J., Wang, F., Li, D., Yang, M., Xing, L., Wei, J., Xiao, H., Yang, Y., Qu, J., Qing, L., Chen, L., Xu, Z., Peng, L., Li, Y., Zheng, H., Chen, F., Huang, K., Jiang, Y., Liu, D., Zhang, Z., Liu, Y., Liu, L. 2020. Treatment of 5 critically ill patients with COVID-19 with convalescent plasma. JAMA—J Am Med Assoc. 323(16), 1582–1589.
44. Ahn, J.Y., Sohn, Y., Lee, S.H., Cho, Y., Hyun, J.H., Baek, Y.J., Jeong, S.J., Kim, J.H., Ku, N.S., Yeom, J.S., Roh, J., Ahn, M.Y., Chin, B.S., Kim, Y.S., Lee, H., Yong, D., Kim, H.O., Kim, S., Choi, J.Y. 2020. Use of convalescent plasma therapy in two covid-19 patients with acute respiratory distress syndrome in Korea. J Korean Med Sci. 35(14), e149.
45. Liu, J.K.H. 2014. The history of monoclonal antibody development—Progress, remaining challenges and future innovations. Ann Med Surg. Elsevier Ltd. 3, 113–116.
46. Singh, S., Kumar, N.K., Dwiwedi, P., Charan, J., Kaur, R., Sidhu, P., Chugh, V.K. 2018. Monoclonal antibodies: A review. Curr Clin Pharmacol. 13(2), 85–99.

47. Gao, Y., Huang, X., Zhu, Y., Lv, Z. 2018. A brief review of monoclonal antibody technology and its representative applications in immunoassays. J Immunoassay Immunochem. Taylor and Francis Inc. 39, 351–364.
48. Todd, P.A., Brogden, R.N. 1989. Muromonab CD3: A review of its pharmacology and therapeutic potential. Drugs. 37(6), 871–899.
49. Shanmugaraj, B., Siriwattananon, K., Wangkanont, K., Phoolcharoen, W. 2020. Perspectives on monoclonal antibody therapy as potential therapeutic intervention for Coronavirus disease-19 (COVID-19). Asian Pac J Allergy Immunol. Allergy and Immunology Society of Thailand. 38, 10–18.
50. Della-Torre, E., Campochiaro, C., Cavalli, G., De Luca, G., Napolitano, A., La Marca, S., Boffini, N., Da Prat, V., Di Terlizzi, G., Lanzillotta, M., Rovere, Querini, P., Ruggeri, A., Landoni, G., Tresoldi, M., Ciceri, F., Zangrillo, A., De Cobelli, F., Dagna, L. 2020. Interleukin-6 blockade with sarilumab in severe COVID-19 pneumonia with systemic hyperinflammation: An open-label cohort study. Ann Rheum Dis. Annrheumdis. 2020–218122, 1–9.
51. Dhimolea, E. 2010. Canakinumab. mAbs. Taylor & Francis. 2, 3–13.
52. Novartis to trial canakinumab to treat Covid-19 pneumonia. 2021. www.clinicaltrialsarena.com/news/novartis-canakinumab-covid-19/.
53. Covid-19: Two antibodies from llamas that can neutralise coronavirus identified—health—Hindustan Times. 2021. www.hindustantimes.com/health/covid-19-two-antibodies-from-llamas-that-can-neutralise-coronavirus-identified/story-ojntN37kFmKXyFaQc6AkJO.html.
54. Sun, X., Wang, T., Cai, D., Hu, Z., Chen, J., Liao, H., Zhi, L., Wei, H., Zhang, Z., Qiu, Y., Wang, J., Wang, A. 2020. Cytokine storm intervention in the early stages of COVID-19 pneumonia. Cytokine Growth Factor Rev. 53, 38–42.
55. Marveh, R., Mohammad, A. M. 2020. Cytokine-targeted therapy in severely ill COVID-19 patients: Options and cautions. EJMO. 4(2), 179–180.
56. Mavrilimumab—Kiniksa | Kiniksa. 2021. www.kiniksa.com/our-pipeline/mavrilimumab/.
57. Kiniksa announces early evidence of treatment response with mavrilimumab in 6 patients with severe COVID-19 pneumonia and hyperinflammation nasdaq: KNSA. 2021. https://pipelinereview.com/index.php/2020040174191/Antibodies/Kiniksa-Announces-Early-Evidence-of-Treatment-Response-with-Mavrilimumab-in-6-Patients-with-Severe-COVID-19-Pneumonia-and-Hyperinflammation.html.
58. Study of the efficacy and safety of ruxolitinib to treat COVID-19 pneumonia—full text view—ClinicalTrials.gov. 2021 . https://clinicaltrials.gov/ct2/show/NCT04331665.
59. Wu, R., Wang, L., Kuo, H.C.D., Shannar, A., Peter, R., Chou, P.J., Li, S., Hudlikar, R., Liu, X., Liu, Z., Poiani, G.J., Amorosa, L., Brunetti, L., Kong, A.N. 2020. An update on current therapeutic drugs treating COVID-19. Curr Pharmacol Reports. 6, 56–70.
60. Vellingiri, B., Jayaramayya, K., Iyer, M., Narayanasamy, A., Govindasamy, V., Giridharan, B., Ganesan, S., Venugopal, A., Venkatesan, D., Ganesan, H., Rajagopalan, K., Rahman, P.K.S.M., Cho, S.G., Kumar, N.S., Subramaniam, M.D. 2020. COVID-19: A promising cure for the global panic. Sci Total Environ. Elsevier B.V. 725, 138277.
61. Lim, J., Jeon, S., Shin, H.Y., Kim, M.J., Seong, Y.M., Lee, W.J., Choe, K.W., Kang, Y.M., Lee, B., Park, S.J. 2019. Case of the index patient who caused tertiary transmission of coronavirus disease 2019 in Korea: The application of lopinavir/ritonavir for the treatment of COVID-19 pneumonia monitored by quantitative RT-PCR. J Korean Med Sci. 35(6), e79.

62. Zhang, Z., Li, X., Zhang, W., Shi, Z.L., Zheng, Z., Wang, T. 2020. Clinical features and treatment of 2019-nCov pneumonia patients in Wuhan: Report of a couple cases. Virol Sin. Science Press. 35(3), 330–336.
63. Caly, L., Druce, J.D., Catton, M.G., Jans, D.A., Wagstaff, K.M. 2020. The FDA-approved drug ivermectin inhibits the replication of SARS-CoV-2 in vitro. Antiviral Res. 178, 104787.
64. Park, S.J., Yu, K.M., Kim, Y.I., Kim, S.M., Kim, E.H., Kim, S.G., Kim, E.J., Casel, M.A.B., Rollon, R., Jang, S.G., Lee, M.H., Chang, J.H., Song, M.S., Jeong, H.W., Choi, Y., Chen, W., Shin, W.J., Jung, J.U., Choi, Y.K. 2020. Antiviral efficacies of FDA-approved drugs against SARS-CoV-2 infection in ferrets. mBio. 11(3), e01114–e01120.
65. Devaux, C.A., Rolain, J.M., Colson, P., Raoult, D. 2020. New insights on the antiviral effects of chloroquine against coronavirus: What to expect for COVID-19? Int J Antimicrob Agents. 55(5), 105938.
66. Keyaerts, E., Li, S., Vijgen, L., Rysman, E., Verbeeck, J., Van, R.M., Maes, P. 2009. Antiviral activity of chloroquine against human coronavirus OC43 infection in newborn mice. Antimicrob Agents Chemother. 53(8), 3416–3421.
67. Kono, M., Tatsumi, K., Imai, A.M., Saito, K., Kuriyama, T., Shirasawa, H. 2008. Inhibition of human coronavirus 229E infection in human epithelial lung cells (L132) by chloroquine: Involvement of p38 MAPK and ERK. Antiviral Res. 77(2), 150–152.
68. Savarino, A., Boelaert, J.R., Cassone, A., Majori, G., Cauda, R. 2003. Effects of chloroquine on viral infections: An old drug against today's diseases? Lancet Infect Dis. Lancet Publishing Group. 3, 722–727.
69. De, W.A.H., Jochmans, D., Posthuma, C.C., Zevenhoven, D.J.C., Van, N.S., Bestebroer, T.M., Van, D.H.B.G., Neyts, J., Snijder, E.J. 2014. Screening of an FDA-approved compound library identifies four small-molecule inhibitors of Middle East respiratory syndrome coronavirus replication in cell culture. Antimicrob Agents Chemother. 58(8), 4875–4884.
70. Mo, Y., Fisher, D. 2016. A review of treatment modalities for middle east respiratory syndrome. J Antimicrob Chemother. 7(12), 3340–3350.
71. Principi, N., Esposito, S. 2020. Chloroquine or hydroxychloroquine for prophylaxis of COVID-19. Lancet Infect Dis. 20(10), 1118.
72. Gautret, P., Lagier, J.C., Parola, P., Hoang, V.T., Meddeb, L., Mailhe, M., Doudier, B., Courjon, J., Giordanengo, V., Vieira, V.E., Dupont, H.T., Honoré, S., Colson, P., Chabrière, E., La, S.B., Rolain, J.M., Brouqui, P., Raoult, D. 2020. Hydroxychloroquine and azithromycin as a treatment of COVID-19: Results of an open-label non-randomized clinical trial. Int J Antimicrob Agents. 56(1), 105949.
73. Rathi, S., Ish, P., Kalantri, A., Kalantri, S. 2020. Hydroxychloroquine prophylaxis for COVID-19 contacts in India. Lancet Infect Dis. Lancet Publishing Group. 20(10), P1118–P1119.
74. FDA cautions against use of hydroxychloroquine or chloroquine for COVID-19 outside of the hospital setting or a clinical trial due to risk of heart rhythm problems | FDA. 2021 . www.fda.gov/drugs/fda-drug-safety-podcasts/fda-cautions-against-use-hydroxychloroquine-or-chloroquine-covid-19-outside-hospital-setting-or.
75. Pastick, K.A., Okafor, E.C., Wang, F., Lofgren, S.M., Skipper, C.P., Nicol, M.R., Pullen, M.F., Rajasingham, R., Mcdonald, E.G., Lee, T.C., Schwartz, I.S., Kelly, L.E., Lother, S.A., Mitjà, O., Letang, E., Abassi, M., Boulware, D.R. 2020. Review: Hydroxychloroquine and chloroquine for treatment of SARS-CoV-2 (COVID-19). Open Forum Infect Dis. 7(4), ofaa130.

76. Owens, B. 2020. Excitement around hydroxychloroquine for treating COVID-19 causes challenges for rheumatology. Lancet Rheumatol. 2(5), e257.
77. France bans hydroxychloroquine to treat covid-19.2021. www.forbes.com/sites/alexledsom/2020/05/27/france-bans-hydroxychloroquine-to-treat-covid-19/?sh=70a64b6a21ab.
78. Coronavirus (COVID-19) update: FDA issues emergency use authorization for potential COVID-19 treatment | FDA. 2021. www.fda.gov/news-events/press-announcements/coronavirus-covid-19-update-fda-issues-emergency-use-authorization-potential-covid-19-treatment.
79. Remdesivir: Uses, interactions, mechanism of action | DrugBank Online91. Solidarity clinical trial for COVID-19 treatments. 2021. https://go.drugbank.com/drugs/DB14761.
80. Cao, B., Wang, Y., Wen, D., Liu, W., Wang, J., Fan, G., Ruan, L., Song, B., Cai, Y., Wei, M., Li, X., Xia, J., Chen, N., Xiang, J., Yu, T., Bai, T., Xie, X., Zhang, L., Li, C., Yuan, Y., Chen, H., Li, H., Huang, H., Tu, S., Gong, F., Liu, Y., Wei, Y., Dong, C., Zhou, F., Gu, X., Xu, J., Liu, Z., Zhang, Y., Li, H., Shang, L., Wang, K., Li, K., Zhou, X., Dong, X., Qu, Z., Lu, S., Hu, X., Ruan, S., Luo, S., Wu, J., Peng, L., Cheng, F., Pan, L., Zou, J., Jia, C., Wang, J., Liu, X., Wang, S., Wu, X., Ge, Q., He, J., Zhan, H., Qiu, F., Guo, L., Huang, C., Jaki, T., Hayden, F.G., Horby, P.W., Zhang, D., Wang, C. 2020. A trial of lopinavir–ritonavir in adults hospitalized with severe covid-19. N Engl J Med. 382(19), 1787–1799.
81. Hung, I.F.N., Lung, K.C., Tso, E.Y.K., Liu, R., Chung, T.W.H., Chu, M.Y., Ng, Y.Y., Lo, J., Chan, J., Tam, A.R., Shum, H.P., Chan, V., Wu, A.K.L., Sin, K.M., Leung, W.S., Law, W.L., Lung, D.C., Sin, S., Yeung, P., Yip, C.C.Y., Zhang, R.R., Fung, A.Y.F., Yan, E.Y.W., Leung, K.H., Ip, J.D., Chu, A.W.H., Chan, W.M., Ng, A.C.K., Lee, R., Fung, K., Yeung, A., Wu, T.C., Chan, J.W.M., Yan, W.W., Chan, W.M., Chan, J.F.W, Lie, A.K.W., Tsang, O.T.Y., Cheng, V.C.C., Que, T.L., Lau, C.S., Chan, K.H., To, K.K.W., Yuen, K.Y. 2020. Triple combination of interferon beta-1b, lopinavir–ritonavir, and ribavirin in the treatment of patients admitted to hospital with COVID-19: An open-label, randomised, phase 2 trial. Lancet. 395(10238), 1695–704.
82. Lopinavir/ritonavir, ribavirin and IFN-beta combination for nCoV treatment—full text view—ClinicalTrials.gov. 2021. https://clinicaltrials.gov/ct2/show/NCT04276688.
83. Lopinavir/ritonavir: A rapid review of effectiveness in COVID-19—CEBM. 2021 . www.cebm.net/covid-19/lopinavir-ritonavir-a-rapid-review-of-the-evidence-for-effectiveness-in-treating-covid/.
84. Tan, E.L.C., Ooi, E.E., Lin, C.Y., Tan, H.C., Ling, A.E., Lim, B., Stanton, L.W. 2004. Inhibition of SARS coronavirus infection in vitro with clinically approved antiviral drugs. Emerg Infect Dis. 10(4), 581–586.
85. Bausch health initiates VIRAZOLE® (Ribavirin for Inhalation Solution, USP) clinical study in patients with COVID-19—Bausch health companies Inc. 2021. https://ir.bauschhealth.com/news-releases/2020/04-13-2020-120008693.
86. Bausch Health initiates VIRAZOLE trial in Covid-19 patients. 2021. www.prnewswire.com/news-releases/bausch-health-initiates-virazole-ribavirin-for-inhalation-solution-usp-clinical-study-in-patients-with-covid-19–301039148.html.
87. Koren, G., King, S., Knowles, S., Phillips, E. 2003. Ribavirin in the treatment of SARS: A new trick for an old drug? CMAJ. 168, 1289–1292.
88. Chakraborty, R., Parvez, S. 2020. COVID-19: An overview of the current pharmacological interventions, vaccines, and clinical trials. Biochem Pharmacol. 180, 114184.

89. Romagnoli, S., Peris, A., Gaudio, A.R.D. 2020. SARS-CoV-2 and COVID-19: From the bench to the bedside. Physioological Rev. 100(4), 1455–1466.
90. Xu, P., Huang, J., Fan, Z., Huang, W., Qi, M., Lin, X., Song, W., Yi, L. 2020. Arbidol/IFN-α2b therapy for patients with corona virus disease 2019: A retrospective multicenter cohort study. Microbes Infect. 22(4–5), 200–205.
91. Blaising, J., Polyak, S.J., Pécheur, E.I. 2014. Arbidol as a broad-spectrum antiviral: An update. Antiviral Res. 107(1), 84–94.
92. Yang, C., Ke, C., Yue, D., Li, W., Hu, Z., Liu, W., Hu, S., Wang, S., Liu, J. 2020. Effectiveness of arbidol for COVID-19 prevention in health professionals. Front Public Heal. 8(249), 1–6.
93. Lin, M.H., Moses, D.C., Hsieh, C.H., Cheng, S.C., Chen, Y.H., Sun, C.Y., Chou, C.Y. 2018. Disulfiram can inhibit MERS and SARS coronavirus papain-like proteases via different modes. Antiviral Res. 150, 155–163.
94. Hu, J.J., Liu, X., Xia, S., Zhang, Z., Zhang, Y., Zhao, J., Ruan, J., Luo, X., Lou, X., Bai, Y., Wang, J., Hollingsworth, L.R., Magupalli, V.G., Zhao, L., Luo, H.R., Kim, J., Lieberman, J., Wu, H. 2020. FDA-approved disulfiram inhibits pyroptosis by blocking gasdermin D pore formation. Nat Immunol. 21(7), 736–745.
95. McCreary, E.K., Pogue, J.M. 2019. Coronavirus disease 2019 treatment: A review of early and emerging options. Open Forum Infect Dis. 7, 1–31.
96. Ahmed, S., Karim, M.M., Ross, A.G., Hossain, M.S., Clemens, J.D., Sumiya, M.K., Phru, C.S., Rahman, M., Zaman, K., Somani, J., Yasmin, R., Hasnat, M.A., Kabir, A., Aziz, A.B., Khan, W.A. 2021. A five-day course of ivermectin for the treatment of COVID-19 may reduce the duration of illness. Int J Infect Dis. 103, 214–216.
97. Gupta, D., Sahoo, A.K., Singh, A. 2020. Ivermectin: Potential candidate for the treatment of Covid-19. Brazilian J Infect Dis. 24(4), 369–371.
98. Heidary, F., Gharebaghi, R. 2020. Ivermectin: A systematic review from antiviral effects to COVID-19 complementary regimen. J Antibiot. Springer US. 73, 593–602.
99. Ono, N., Tatsuo, H., Hidaka, Y., Aoki, T., Minagawa, H., Yanagi, Y. 2001. Measles viruses on throat swabs from measles patients use signaling lymphocytic activation molecule (CDw150) but not CD46 as a cellular receptor. J Virol. 75(9), 4399–4401.
100. Zhang, H., Wang, C.Y., Zhou, P., Yue, H., Du, R. 2020. Histopathologic changes and SARS-CoV-2 immunostaining in the lung of a patient with COVID-19. Ann Intern Med. 172(9), 629–632.
101. Schmith, V.D., Zhou, J., Lohmer, L.R.L. 2020. The approved dose of ivermectin alone is not the ideal dose for the treatment of COVID-19. Clin Pharmacol Ther. 108(4), 762–765.
102. Gyselinck, I., Janssens, W., Verhamme, P., Vos, R. 2021. Rationale for azithromycin in COVID-19: An overview of existing evidence. BMJ Open Respir Res. 8, 1–10.
103. Zimmermann, P., Ziesenitz, V.C., Curtis, N., Ritz, N. 2018. The immunomodulatory effects of macrolides-A systematic review of the underlying mechanisms. Front Immunol. 9, 302.
104. Cusinato, J., Cau, Y., Calvani, A.M., Mori, M. 2021. Repurposing drugs for the management of COVID-19. Expert Opin Ther Pat. Taylor & Francis. 31, 295–307.
105. Fantini, J., Chahinian, H., Yahi, N. 2020. Synergistic antiviral effect of hydroxychloroquine and azithromycin in combination against SARS-CoV-2: What molecular dynamics studies of virus-host interactions reveal. Int J Antimicrob Agents. 56(2), 106020.
106. Ulrich, H., Pillat, M.M. 2020. CD147 as a target for COVID-19 treatment: Suggested effects of azithromycin and stem cell engagement. Stem Cell Rev Reports. 16(3), 434–440.

107. Akhoundimeybodi, Z., Mousavinasab, S.R., Owlia, M.B., Owlia, S. 2021. Efficacy and and safety doxycycline in treating COVID-19 positive patients: A pilot clinical study. Pakistan J Med Heal Sci. 15(1), 610–614.
108. Malek, A.E., Granwehr, B.P., Kontoyiannis, D.P. 2020. Doxycycline as a potential partner of COVID-19 therapies. IDCases. 21, e00864.
109. Gendrot, M., Andreani, J., Jardot, P., Hutter, S., Delandre, O., Boxberger, M., Mosnier, J., Le Bideau, M., Duflot, I., Fonta, I., Rolland, C., Bogreau, H., La Scola, B., Pradines, B. 2020. In vitro antiviral activity of doxycycline against SARS-CoV-2. Molecules. 25(21), 1–10.
110. Stockman, L.J., Bellamy, R., Garner, P. 2006. SARS: Systematic review of treatment effects. PLoS Med. 3(9), 1525–1531.
111. Shang, L., Zhao, J., Hu, Y., Du, R., Cao, B. 2020. On the use of corticosteroids for 2019-nCoV pneumonia. Lancet. 395(10225), 683–684.
112. de Faria, C.R.C., Campos, C.F., Ziegler, S. F.L.F., Marques, M.P. C., Laviano, A., Mota, J.F. 2021. Dietary recommendations during the COVID-19 pandemic. Kompass Nutr Diet. 1(1), 3–7.
113. Polak, E., Stępień, A.E., Gol, O., Tabarkiewicz, J. 2021. Potential immunomodulatory effects from consumption of nutrients in whole foods and supplements on the frequency and course of infection: Preliminary results. Nutrients. 13(4), 1157, 1–11.
114. Milani, G.P., Macchi, M., Guz-Mark, A. 2021. Vitamin c in the treatment of covid-19. Nutrients. 13(4), 1–10.
115. Khan, A.H, Nasir, N., Nasir, N., Maha, Q., Rehman, R. 2021. Vitamin D and COVID-19: Is there a role? J Diabetes Metab Disord. 20, 931–938.
116. Turrubiates-Hernández, F.J., Sánchez-Zuno, G.A., Gonzalez-Estevez, G., Hernandez-Bello, J., Macedo-Ojeda, G., José, M.V.F. 2019. Potential immunomodulatory effects of vitamin D in the prevention of severe coronavirus disease 2019: An ally for Latin America (Review). Int J Mol Med. 47(4), 1–15.
117. Acosta-Elias, J., Espinosa-Tanguma, R. 2020. The folate concentration and/or folic acid metabolites in Plasma as factor for COVID-19 infection. Front Pharmacol. 11, 1062, 1–4.
118. Sundararaman, A., Ray, M., Ravindra, P.V., Halami, P.M. 2020. Role of probiotics to combat viral infections with emphasis on COVID-19. Appl Microbiol Biotechnol. 104(19), 8089–8104.
119. Yan, F., Polk, D.B. 2011. Probiotics and immune health. Curr Opin Gastroenterol. 27, 496–501.
120. Kumar, P., Kumar, M., Bedi, O., Gupta, M., Kumar, S., Jaiswal, G., Rahi, V., Yedke, N.G., Bijalwan, A., Sharma, S., Jamwal, S. 2021. Role of vitamins and minerals as immunity boosters in COVID-19. Inflammopharmacology. 2, 0123456789.
121. Prasad, A.S. 2008. Zinc in human health: Effect of zinc on immune cells. Mol. Med. 14, 353–357.
122. Saeidi, A., Tayebi, S.M., To-aj, O., Karimi, N., Kamankesh, S., Niazi, S., Khosravi, A., Khademosharie, M., Soltani, M., Johnson, K.E., Rashid, H., Laher, I., Hackney, A.C., Zouhal, H. 2021. Physical activity and natural products and minerals in the SARS-CoV-2 pandemic: An update. Ann Appl Sport Sci. 9(1), 1–26.
123. Alagawany, M., Attia, Y.A., Farag, M.R., Elnesr, S.S., Nagadi, S.A., Shafi, M.E., Khafaga, A.F., Ohran, H., Alaqil, A.A., Abd El-Hack, M.E. 2021. The strategy of boosting the immune system under the COVID-19 pandemic. Front Vet Sci. 7, 570784.
124. Akhtar, S., Das, J.K., Ismail, T., Wahid, M., Saeed, W., Bhutta, Z.A. 2021. Nutritional perspectives for the prevention and mitigation of COVID-19. Nutr Rev. 79(3), 289–300.

125. Moderna COVID-19 vaccine I. 2021 . www.fda.gov/emergency-preparedness-and-response/coronavirus-disease-2019-covid-19/moderna-covid-19-vaccine.
126. Safety and immunogenicity study of 2019-nCoV vaccine (mRNA-1273) for prophylaxis of SARS-CoV-2 infection (COVID-19)—full text view—ClinicalTrials.gov. 2021 . www.clinicaltrials.gov/ct2/show/NCT04283461.
127. Anderson, E.J., Rouphael, N.G., Widge, A.T., Jackson, L.A., Roberts, P.C., Makhene, M., Chappell, J.D., Denison, M.R., Stevens, L.J., Pruijssers, A.J., McDermott, A.B., Flach, B., Lin, B.C., Doria-Rose, N.A., O'Dell, S., Schmidt, S.D., Corbett, K.S., Swanson, P.A., Padilla, M., Neuzil, K.M., Bennett, H., Leav, B., Makowski, M., Albert, J., Cross, K., Edara, V.V., Floyd, K., Suthar, M.S., Martinez, D.R., Baric, R., Buchanan, W., Luke, C.J., Phadke, V.K., Rostad, C.A., Ledgerwood, J.E., Graham, B.S., Beigel, J.H. 2020. Safety and immunogenicity of SARS-CoV-2 mRNA-1273 vaccine in older adults. N Engl J Med. 383(25), 2427–2438.
128. Developing therapeutics and vaccines for coronaviruses I NIH: National institute of allergy and infectious diseases. 2021 . www.niaid.nih.gov/diseases-conditions/coronaviruses-therapeutics-vaccines.
129. Beigel, J.H., Voell, J., Kumar, P., Raviprakash, K., Wu, H., Jiao, J.A., Sullivan, E., Luke, T., Davey, R.T. 2018. Safety and tolerability of a novel, polyclonal human anti-MERS coronavirus antibody produced from transchromosomic cattle: A phase 1 randomised, double-blind, single-dose-escalation study. Lancet Infect Dis. 18(4), 410–418.
130. Experimental MERS treatments enter clinical trial I NIH: National institute of allergy and infectious diseases. 2021 . www.niaid.nih.gov/news-events/experimental-mers-treatments-enter-clinical-trial.
131. https://covid19.trackvaccines.org/vaccines/17/.
132. AstraZeneca and Oxford University announce landmark agreement for COVID-19 vaccine. 2021. www.astrazeneca.com/media-centre/press-releases/2020/astrazeneca-and-oxford-university-announce-landmark-agreement-for-covid-19-vaccine.html.
133. AstraZeneca COVID-19 vaccine Vaxzevria authorised for emergency use in Japan. 2021. www.astrazeneca.com/media-centre/press-releases/2021/astrazeneca-covid-19-vaccine-vaxzevria-authorised-for-emergency-use-in-japan.html.
134. Voysey, M., Clemens, S.A.C., Madhi, S.A., Weckx, L.Y., Folegatti, P.M., Aley, P.K., Angus, B., Baillie, V.L., Barnabas, S.L., Bhorat, Q.E., Bibi, S., Briner, C., Cicconi, P., Collins, A.M., Colin-Jones, R., Cutland, C.L., Darton, T.C., Dheda, K., Duncan, C.J.A., Emary, K.R.W., Ewer, K.J., Fairlie, L., Faust, S.N., Feng, S., Ferreira, D.M., Finn, A., Goodman, A.L., Green, C.M., Green, C.A., Heath, P.T., Hill, C., Hill, H., Hirsch, I., Hodgson, S.H.C., Izu, A., Jackson, S., Jenkin, D., Joe, C.C.D. 2021. Safety and efficacy of the ChAdOx1 nCoV-19 vaccine (AZD1222) against SARS-CoV-2: An interim analysis of four randomised controlled trials in Brazil, South Africa, and the UK. Lancet. 397(10269), 99–111.
135. COVID-19 vaccine AstraZeneca effective against Delta ('Indian') variant. 2021. www.astrazeneca.com/media-centre/press-releases/2021/covid-19-vaccine-astrazeneca-effective-against-delta-indian-variant.html.
136. Serum institute of India—ChAdOx1 nCoV-19 corona virus vaccine (Recombinant)—COVISHIELD. 2021. www.seruminstitute.com/product_covishield.php.
137. International partnership progresses UQ COVID-19 vaccine project—UQ News—The University of Queensland, Australia. https://www.uq.edu.au/news/article/2020/04/international-partnership-progresses-uq-covid-19-vaccine-project
138. Johnson & Johnson announces collaboration with U.S. department of health & human services to accelerate development of a potential novel coronavirus vaccine I Johnson & Johnson. 2021. www.uq.edu.au/news/article/2020/04/international-partnership-progresses-uq-covid-19-vaccine-project.

139. Sanofi announces it will work with HHS to develop coronavirus vaccine. 2021 . www.scientificamerican.com/article/sanofi-announces-it-will-work-with-hhs-to-develop-coronavirus-vaccine/.
140. Coronavirus vaccine: Pfizer, BioNTech enter race for covid drug—Bloomberg. 2021 . www.bloomberg.com/news/articles/2020-04-22/pfizer-biontech-coronavirus-vaccine-trial-to-start-in-germany.
141. Johnson & Johnson COVID-19 vaccine authorized by U.S. FDA for emergency use | Johnson & Johnson. 2021. www.jnj.com/johnson-johnson-covid-19-vaccine-authorized-by-u-s-fda-for-emergency-usefirst-single-shot-vaccine-in-fight-against-global-pandemic.
142. Ella, R., Reddy, S., Jogdand, H., Sarangi, V., Ganneru, B., Prasad, S., Das, D., Raju, D., Praturi, U., Sapkal, G., Yadav, P., Reddy, P., Verma, S., Singh, C., Redkar, S.V., Gillurkar, C.S., Kushwaha, J.S., Mohapatra, S., Bhate, A., Rai, S., Panda, S., Abraham, P., Gupta, N., Ella, K., Bhargava, B., Vadrevu, K.M. 2021. Safety and immunogenicity of an inactivated SARS-CoV-2 vaccine, BBV152: Interim results from a double-blind, randomised, multicentre, phase 2 trial, and 3-month follow-up of a double-blind, randomised phase 1 trial. Lancet Infect Dis. 9(13), 14–18.
143. Whole-virion inactivated SARS-CoV-2 vaccine (BBV152) for COVID-19 in healthy volunteers. 2021 . https://clinicaltrials.gov/ct2/show/NCT04471519.
144. COVAXIN—India's first indigenous covid-19 Vaccine | Bharat Biotech. 2021 . www.bharatbiotech.com/covaxin.html.
145. Ganneru, B., Jogdand, H., Daram, V.K., Das, D., Molugu, N.R., Prasad, S.D., Kannappa, S.V., Ella, K.M., Ravikrishnan, R., Awasthi, A., Jose, J., Rao, P., Kumar, D., Ella, R., Abraham, P., Yadav, P.D., Sapkal, G.N., Shete-Aich, A., Desphande, G., Mohandas, S., Vadrevu, K.M. 2021. Th1 skewed immune response of whole virion inactivated SARS CoV 2 vaccine and its safety evaluation. *iScience*. 24(4), 102298.
146. Covaxin phase 3 trial data shows 77.8% efficacy, gets approval from DCGI's expert panel. 2021 . https://news.abplive.com/news/india/covaxin-phase-3-trial-data-shows-77–8-efficacy-gets-approval-from-dcgi-s-expert-panel-1465042.
147. Yadav, P.D., Sapkal, G.N., Ella, R., Sahay, R.R., Nyayanit, D.A., Patil, D.Y., Deshpande, G., Shete, A.M., Gupta, N., Mohan, V.K., Abraham, P., Panda, S., Bhargava, B., Neutralization against B.1.351 and B.1.617.2 with sera of COVID-19 recovered cases and vaccinees of BBV152. bioRxiv i. 2021 . https://doi.org/10.1101/2021.04.23.441101.
148. Yadav, P.D., Ella, R., Kumar, S., Patil, D.R., Mohandas, S., Shete, A.M., Vadrevu, K.M., Bhati, G., Sapkal, G., Kaushal, H., Patil, S., Jain, R., Deshpande, G., Gupta, N., Agarwal, K., Gokhale, M., Mathapati, B., Metkari, S., Mote, C., Nyayanit, D., Patil, D.Y., Sai Prasad, B.S., Suryawanshi, A., Kadam, M., Kumar, A., Daigude, S., Gopale, S., Majumdar, T., Mali, D., Sarkale, P., Baradkar, S., Gawande, P., Joshi, Y., Fulari, S., Dighe, H., Sharma, S., Gunjikar, R., Kumar, A., Kalele, K., Srinivas, V.K., Gangakhedkar, R.R., Ella, K.M., Abraham, P., Panda, S., Bhargava, B. 2021. Immunogenicity and protective efficacy of inactivated SARS-CoV-2 vaccine candidate, BBV152 in rhesus macaques. Nat Commun. 12(1), 1–11.
149. Mohandas, S., Yadav, P.D., Shete-Aich, A., Abraham, P., Vadrevu, K.M., Sapkal, G., Mote, C., Nyayanit, D., Gupta, N., Srinivas, V.K., Kadam, M., Kumar, A., Majumdar, T., Jain, R., Deshpande, G., Patil, S., Sarkale, P., Patil, D., Ella, R., Prasad, S.D., Sharma, S., Ella, K.M., Panda, S., Bhargava, B. 2021. Immunogenicity and protective efficacy of BBV152, whole virion inactivated SARS-CoV-2 vaccine candidates in the Syrian hamster model. iScience. 24(2), 102054.

150. Yadav, P.D., Sapkal, G.N., Abraham, P., Ella, R., Deshpande, G., Patil, D.Y., Nyayanit, D.A., Gupta, N., Sahay, R.R., Shete, A.M., Panda, S., Bhargava, B., Mohan, V.K., Neutralization of variant under investigation B.1.617 with sera of BBV152 vaccinees. Clin Infect Dis. 2021. https://academic.oup.com/cid/advancearticle/doi/10.1093/cid/ciab411/6271524.
151. The ministry of health of the Russian federation approved the use of the Sputnik V vaccine for people over 60 years old. 2021. https://minzdrav.gov.ru/news/2020/12/26/15765-minzdrav-rf-odobril-primenenie-vaktsiny-sputnik-v-dlya-lyudey-starshe-60-let.
152. Sputnik, V. 2021. About vaccine | Official website vaccine against COVID-1 9 https://sputnikvaccine.com/about-vaccine/.
153. Gamalei, N.F. 2021. NITsEM them . www.gamaleya.org/.
154. Logunov, D.Y., Dolzhikova, I.V., Shcheblyakov, D.V., Tukhvatulin, A.I., Zubkova, O.V., Dzharullaeva, A.S., Kovyrshina, A.V., Lubenets, N.L., Grousova, D.M., Erokhova, A.S., Botikov, A.G., Izhaeva, F.M., Popova, O., Ozharovskaya, T.A., Esmagambetov, I.B., Favorskaya, I.A., Zrelkin, D.I., Voronina, D.V., Shcherbinin, D.N., Semikhin, A.S., Simakova, Y.V., Tokarskaya, E.A., Egorova, D.A., Shmarov, M.M., Nikitenko, N.A., Gushchin, V.A., Smolyarchuk, E.A., Zyryanov, S.K., Borisevich, S.V., Naroditsky, B.S., Gintsburg, A.L. 2021. Safety and efficacy of an rAd26 and rAd5 vector-based heterologous prime-boost COVID-19 vaccine: An interim analysis of a randomised controlled phase 3 trial in Russia. Lancet. 397(10275), 671–681.
155. Novavax COVID-19 vaccine demonstrates 90% overall efficacy and 100% protection against moderate and severe disease in PREVENT-19 phase 3 trial | Novavax Inc.—IR Site. 2021. https://ir.novavax.com/news-releases/news-release-details/novavax-covid-19-vaccine-demonstrates-90-overall-efficacy-and.
156. A study to evaluate the efficacy, immune response, and safety of a COVID-19 vaccine in adults ≥ 18 years with a pediatric expansion in adolescents (12–17 years) at risk for SARS-CoV-2—full text view—ClinicalTrials.gov. 2021 . https://clinicaltrials.gov/ct2/show/NCT04611802.
157. Evaluation of the safety and immunogenicity of a SARS-CoV-2 rS nanoparticle vaccine with/without matrix-M adjuvant—full text view—ClinicalTrials.gov. 2021 . https://clinicaltrials.gov/ct2/show/NCT04368988.
158. A study looking at the effectiveness, immune response, and safety of a COVID-19 vaccine in adults in the United Kingdom—full text view—ClinicalTrials. 2021 . https://clinicaltrials.gov/ct2/show/NCT04583995.
159. WHO lists additional COVID-19 vaccine for emergency use and issues interim policy recommendations. 2021 . www.who.int/news/item/07-05-2021-who-lists-additional-covid-19-vaccine-for-emergency-use-and-issues-interim-policy-recommendations.
160. The sinopharm COVID-19 vaccine: What you need to know. 2021 . www.who.int/news-room/feature-stories/detail/the-sinopharm-covid-19-vaccine-what-you-need-to-know.
161. Polack, F.P., Thomas, S.J., Kitchin, N., Absalon, J., Gurtman, A., Lockhart, S., Perez, J.L., Pérez Marc, G., Moreira, E.D., Zerbini, C., Bailey, R., Swanson, K.A., Roychoudhury, S., Koury, K., Li, P., Kalina, W.V., Cooper, D., Frenck, R.W., Hammitt, L.L., Türeci, Ö., Nell, H., Schaefer, A., Ünal, S., Tresnan, D.B., Mather, S., Dormitzer, P.R., Şahin, U., Jansen, K.U., Gruber, W.C. 2020. Safety and efficacy of the BNT162b2 mRNA covid-19 vaccine. N Engl J Med. 383(27), 2603–2615.

162. Study to describe the safety, tolerability, immunogenicity, and efficacy of RNA vaccine candidates against COVID-19 in healthy individuals—full text view—ClinicalTrials.gov. 2021 . https://clinicaltrials.gov/ct2/show/NCT04368728.
163. Frenck, R.W., Klein, N.P., Kitchin, N., Gurtman, A., Absalon, J., Lockhart, S., Perez, J.L., Walter, E.B., Senders, S., Bailey, R., Swanson, K.A., Ma, H., Xu, X., Koury, K., Kalina, W.V., Cooper, D., Jennings, T., Brandon, D.M., Thomas, S.J., Türeci, Ö., Tresnan, D.B., Mather, S., Dormitzer, P.R., Şahin, U., Jansen, K.U., Gruber, W.C. 2021. C4591001 Clinical trial group. Safety, immunogenicity, and efficacy of the BNT162b2 covid-19 vaccine in adolescents. N Engl J Med. 385(3), 239–250.
164. COVID-19: Tests for 'miracle cure' herb Artemisia begin | Africa | DW | 15.05.2020 . www.dw.com/en/covid-19-tests-for-miracle-cure-herb-artemisia-begin/a-53442366.
165. Ashwagandha takes lead in IIT-Delhi study to be Covid-19 warrior—The Hindu BusinessLine. 2021 . www.thehindubusinessline.com/news/science/ashwagandha-takes-lead-in-iit-delhi-study-to-be-covid-19-warrior/article31621996.ece.
166. https://clinicaltrials.gov/ct2/show/record/NCT04382040.
167. AYUSH releases immunity-boosting measures for self-care during COVID-19 pandemic: Here is what you can do. 2021 . www.dnaindia.com/health/report-ayush-releases-immunity-boosting-measures-for-self-care-during-covid-19-pandemic-here-is-what-you-can-do-2820584.
168. The role of resistant starch in COVID-19 infection—full text view—ClinicalTrials.gov. 2021 . https://clinicaltrials.gov/ct2/show/results/NCT04342689?view=results.
169. Efficacy of natural honey treatment in patients with novel coronavirus—full text view—ClinicalTrials.gov. 2021. https://clinicaltrials.gov/ct2/show/results/NCT04323345.
170. Omega-3 fatty acids and health effects—full text view—ClinicalTrials.gov. 2021. www.clinicaltrials.gov/ct2/show/NCT01034423.
171. Vitamin C infusion for the treatment of severe 2019-nCoV infected pneumonia—full text view—ClinicalTrials.gov. 2021. https://clinicaltrials.gov/ct2/show/NCT04264533.
172. A study of hydroxychloroquine, vitamin C, vitamin D, and zinc for the prevention of COVID-19 infection—full text view—ClinicalTrials.gov. 2021 . https://clinicaltrials.gov/ct2/show/NCT04335084.
173. Hydroxychloroquine in patients with newly diagnosed COVID-19 compared to standard of care—full text view—ClinicalTrials.gov. https://classic.clinicaltrials.gov/ct2/show/NCT04334967
174. Norwegian coronavirus disease 2019 study—full text view—ClinicalTrials.gov. https://classic.clinicaltrials.gov/ct2/show/NCT04316377
175. Clinical Trial of the Use of Tocilizumab for Treatment of SARS-CoV-2 Infection (COVID-19)—full text view—ClinicalTrials.gov. https://www.cdek.liu.edu/trial/NCT04332094
176. Comparison of lopinavir/ritonavir or hydroxychloroquine in patients with mild coronavirus disease (COVID-19)—full text view—ClinicalTrials.gov. https://classic.clinicaltrials.gov/ct2/show/NCT04307693
177. Safety and efficacy of hydroxychloroquine associated with azithromycin in SARS-CoV2 virus (coalition covid-19 Brasil II)—full text view—ClinicalTrials.gov. https://classic.clinicaltrials.gov/ct2/show/NCT04321278
178. Chloroquine diphosphate for the treatment of severe acute respiratory syndrome secondary to SARS-CoV2—full text view—ClinicalTrials.gov. https://classic.clinicaltrials.gov/ct2/show/NCT04323527

179. Study to evaluate the safety and antiviral activity of remdesivir (GS-5734™) in participants with severe coronavirus disease (COVID-19)—full text view—Clinical Trials.gov. https://classic.clinicaltrials.gov/ct2/show/NCT04292899
180. Hadad, D J., et al. 2004. Mycobacteraemia among HIV-1-infected patients in São Paulo, Brazil: 1995 to 1998. Epidemiol Infect. 132(1), 151–155.
181. Adaptive COVID-19 treatment trial (ACTT)—full text view—ClinicalTrials.gov. https://classic.clinicaltrials.gov/ct2/show/NCT04280705
182. Trial of treatments for COVID-19 in hospitalized adults—full text view—ClinicalTrials.gov. https://classic.clinicaltrials.gov/ct2/show/NCT04315948
183. A trial of remdesivir in adults with mild and moderate COVID-19—full text view—ClinicalTrials.gov. [Online]. https://clinicaltrials.gov/ct2/show/NCT04252664. [Accessed: 21-Apr-2020].
184. A trial of remdesivir in adults with severe COVID-19—full text view—ClinicalTrials.gov. [Online]. https://clinicaltrials.gov/ct2/show/NCT04257656. [Accessed: 21-Apr-2020].
185. Lopinavir/ritonavir, ribavirin and IFN-beta combination for nCoV treatment—full text view—ClinicalTrials.gov. https://classic.clinicaltrials.gov/ct2/show/NCT04276688
186. The efficacy of lopinavir plus ritonavir and arbidol against novel coronavirus infection—full text view—ClinicalTrials.gov. https://classic.clinicaltrials.gov/ct2/show/NCT04252885

10 Exploration of Nanotherapeutics in COVID-19

Nikhat J. Siddiqi, Sabiha Fatima, Sooad K. AlDaihan, Hanan A. Balto and Bechan Sharma

10.1 INTRODUCTION

Viruses cause about one-third of death due to infectious diseases. Viruses causing human immunodeficiency virus (HIV) infections and acute respiratory infections are known to cause worldwide mortality (Brachman P.S., 2003; Morens et al., 2004). Diseases of the respiratory tract, especially those affecting the lower respiratory tract, are caused by viruses like influenza viruses, respiratory syncytial viruses, rhinoviruses, etc (Gunathilake et al., 2021). Severe acute respiratory syndrome coronavirus-2 (SARS) is the most recent infectious disease caused by a coronavirus (Shan et al., 2020). It is caused by the virus belonging to the family Coronaviridae or coronaviruses. The family Coronaviridae comprises of two subfamilies: Torovirinae and Coronavirinae. In humans Coronaviruses cause respiratory diseases like the common cold, Middle East respiratory syndrome, severe acute respiratory syndrome and COVID-19 (Pal et al., 2020). In recent times, the major pandemic of the millennium was caused by coronavirus (Drosten et al., 2003). The first occurrence of SARS coronavirus infection caused a fatality of about 10% (Cheng et al., 2007). In 2012, a global epidemic was caused by Middle East respiratory syndrome which affected 27 countries worldwide (Bchetnia et al., 2020).

A unique coronavirus causing pneumonia was identified from Wuhan, China, in December 2019. Ihe following March 2020, COVID-19 was declared a pandemic by WHO (Gautret et al., 2020). This virus was named as SARS-CoV-2 by the International Committee for Virus Taxonomy (Gunathilake et al., 2021).

On 11 February, the International Committee for Virus Taxonomy (ICTV)'s Coronavirus Research Group (CGS) named the virus as SARS-CoV-2 based on the study of its evolutionary past of the and the pathogen causing severe acute respiratory syndrome (SARS) (Gunathilake et al., 2021). The World Health Organization (WHO) announced that the coronavirus outbreak (COVID-19) was caused by SARS-CoV-2. Coronavirus (COVID-19) has caused more than three million deaths worldwide (Seah and Agrawal, 2020). The sequencing of SARCS-CoV-2 genome has shown that its genes are identical to those of other

DOI: 10.1201/9781003324911-15

coronaviruses that cause respiratory illnesses such as SARS-CoV (Gunathilake et al., 2021).

Some of the common symptoms of COVID-19 include fever, dry cough, fatigue, headache, loss of appetite, etc. (Gunathilake et al., 2021).

10.2 STRUCTURE OF SEVERE ACUTE RESPIRATORY SYNDROME CORONAVIRUS 2 (SARS-CoV-2)

Coronaviruses belongs to the subfamily *Coronavirinae* in the family of Coronaviridae and the subfamily contains four genera: *Alphacoronavirus*, *Betacoronavirus*, *Gammacoronavirus* and *Deltacoronavirus*. The genome of CoVs (27–32 kb) is a single-stranded positive-sense RNA (+ssRNA), which is larger than any other RNA viruses. The nucleocapsid protein (N) is formed in the capsid outside the genome. The genome is further packed by an envelope which is associated with three structural proteins: membrane protein (M), spike protein (S) and envelope protein (E) (Brian and Baric, 2005). The genome size of SARS-CoV-2 was reported to be approximately 29.9 kb (Lu et al., 2020). The S protein helps infection by facilitating SARS-CoV-2 to determine the angiotensin-converting enzyme 2 (ACE2) and thereby invade the host (Figure 10.1).

10.3 INFECTION

Viruses like influenza, respiratory syncytial and parainfluenza viruses infect commonly through the respiratory pathways. These virulent pathogens first attack the upper respiratory tract. As the diseases progresses, they enter the lower breathing airways (Al-Halifa et al., 2019). They are mostly transmitted as airborne infections through direct contact or droplets and aerosols. They then undergo replication in the respiratory tract, which is accompanied by the appearance of clinical

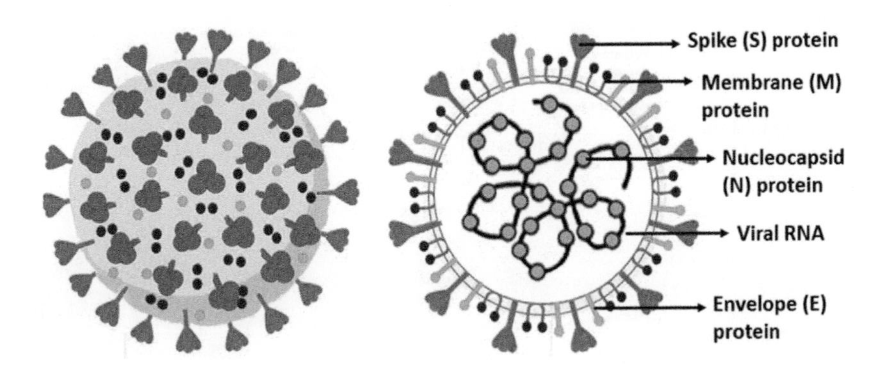

FIGURE 10.1 Structure of severe acute respiratory syndrome coronavirus-2.

symptoms like fever, dyspnea, cough, bronchiolitis and pneumonia (Arruda et al., 2006; Boncristiani et al., 2009).

The virus binds to a receptor on the host cell, where it merges with the cell membrane (Rothan and Byrareddy, 2020). The human cell receptor is the angiotensin-converting enzyme 2, known as ACE2 (Khan et al., 2020). The transmembrane serine protease 2 causes the viral entry into the cell. Inside the cell, the viral particles are taken into endosomes. The low pH in the endosomes uncoats the viral particles. The viral genome is then released for protein synthesis. This is followed by the synthesis of viral proteins and RNA, their assembly into infectious viral particles and their release outside the cell (Gunathilake et al., 2021). The events of the infection are depicted in Figure 10.2.

Once the lung epithelium is damaged, the immune response releases proinflammatory cytokines (cytokine storms), causing acute respiratory distress syndrome and multiple organ failure (Li et al., 2020; Petrosillo et al., 2020). The SARS-CoV-2 causes an increased mucous discharge in the respiratory system. The mucus clogs the alveoli and results in prevention of blood oxygenation. The virus can also spread to the digestive system and other organs, viz., kidney and liver. It can invade every tissue that expresses angiotensin-converting enzyme-2 receptor (Gavriatopoulou et al., 2020; Glebov OO, 2020; Ullah et al., 2020).

Nanomaterials and advanced nanotechnology provide a platform for better understanding of interaction of the virus with its host. This in turn provides leads to the development of antiviral treatment. Engineered nanoparticles are used to prevent the virus from entering the host cell and to inactivate the virus. For the past several years metallic nanoparticles like gold, silver, iron, etc., are being used in viral diagnostics

Metal nanoparticles like gold, silver, zinc and copper are used in magnetic immunoassay and viral diagnostics. This has encouraged the researchers to consider the possibility of using nanoparticles for detection/treatment and prevention of COVID-19.

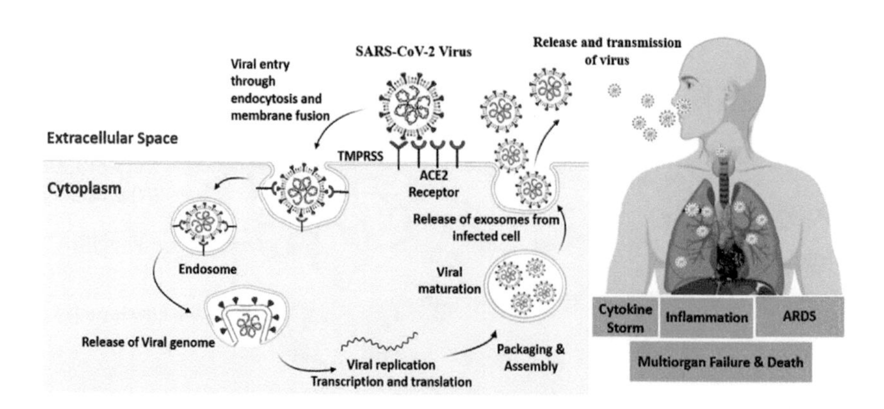

FIGURE 10.2 Life cycle of severe acute respiratory syndrome coronavirus 2.

10.4 NANOTHERAPY

Nanostructured materials/nanoparticles (NPs) are materials with at least one nanometric dimension usually less than 100 nm. They can be organic, inorganic, biomaterial-based and carbon-based (Melchor-Martínez et al., 2021). Their unique physicochemical properties like chemical reactivity, size-dependent transport, biocompatibility and reduced toxicity can be exploited in the treatment/management of COVID-19.

Both the NPs and virus possess nanoscale level dimensions (Figure 10.3). Therefore, NPs are attractive candidates for vaccine development and immune engineering. Due to their extreme small size, NPs are capable of binding, encapsulating the virus. This helps in detection, treatment and prevention of viral disease (Gregory et al., 2013). Nanomaterials kill the virus either by direct interaction or by inducing stimuli which can kill the virus. The unique surface properties of NP makes them attractive antiviral agents (Seifi and Reza, 2021). Nano diagnostics, nanotherapeutics, nanovaccination, etc., are some of applications of nanotechnology which can be exploited to fight the outbreak of COVID-19 (Chan WCW., 2020) (Table 10.1).

The antibacterial, antifungal, antiparasitic and antiviral properties of some NPs have been used in medicine. These NPs include soft (polymeric nanoparticles), lipids (lipid-solid nanoparticles, nanostructured lipid carriers, liposomes) and metal nanoparticles. Drug based nanoparticles have been tested in clinical trials for cancer, neurodegenerative, cardiac and infectious diseases (Kupferschmidt and Cohen, 2020). Nanotechnology offers new avenues for antiviral therapy for new viruses (Mainardes and Diedrich, 2020).

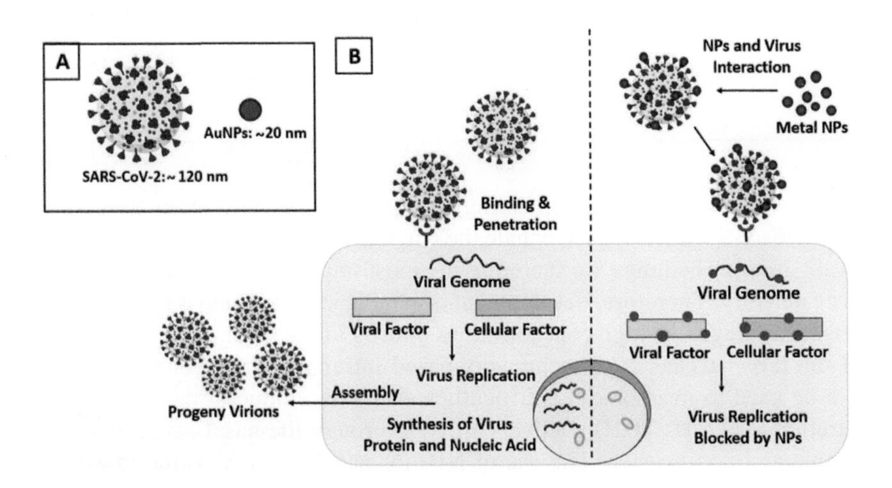

FIGURE 10.3 Comparison between the average size of SARS-CoV-2 and nanoparticles (A). Nanoparticles can interfere with viral replication (B).

TABLE 10.1
Some of Nanotechnology-Based Therapeutics Used in Different Types of Coronavirus Infections

	Types of nanotechnology-based therapeutics	The purpose of use	References
Severe acute respiratory syndrome coronavirus	Gold nanoparticle	To act as an adjuvant for S protein vaccine	Sekimukai et al., 2020.
Severe acute respiratory syndrome coronavirus	Alum adjuvanted nanoparticles of coronavirus S protein	To promote the neutralising coronavirus antibody response in mice	Coleman et al., 2014
Severe acute respiratory syndrome coronavirus	SARS DNA in a dynamic pci-S/ polyethylenimine nanoparticles complex (DNA vaccine)	To act as a potent mucosal immunostimulant for intranasal immunisation	Shim et al., 2010
Severe acute respiratory syndrome coronavirus	Dendritic cell targeting chitosan nanoparticles	To develop nasal route vaccine delivery system using dendritic cell targeting	Raghuwanshi et al., 2012
Severe acute respiratory syndrome coronavirus	Modified version of the peptide nanoparticle (subunit vaccine)	To obtain repeated antigen display system	Pimentel et al., 2009
Severe acute respiratory syndrome coronavirus 2	SaRNA encapsulated lipid nanoparticle (RNA vaccine)	To encapsulate self-amplifying RNA within lipid nanoparticle as a vaccine	McKay et al., 2020

The respiratory tract including the lungs are the main targets of SARS-CoV-2. Other targets include the gut, kidney and vasculatures (Lukassen et al., 2020). The lungs are therefore the most important organ for COVID-19 drug delivery. Therefore, development of inhalable NPs would circumvent side effects arising from high concentrations of drug in the serum or inaccessibility of the target tissue. Nanotechnology-based intranasal drug delivery systems can be used to overcome the difficulties encountered due to mucosal administration (Yang D., 2021). Delivery of NPs through the nasal cavity would be simple and noninvasive. The use of NPs for nasal delivery would make NPs rapidly absorptive and also reduce the cost of the treatment (Costantino et al.,

2007). Biocompatible nanomaterials such as nanosheets made of boron nitride oxide can increase the adsorption of drugs to viral protein. This would aid the drug to rapidly diffuse to the viral protein and aid in drug-virus interaction (Duverger et al., 2021).

The strategy to combat SARS-CoV-2 by nanoparticles could involve mechanisms that prevent the entry of the virus in the host. The blockage of the viral surface proteins may cause virus inactivation and thus lead to reduced viral internalisation. Metallic nanoparticles, viz., titanium, silver, gold and zinc have been demonstrated to prevent viral attachment to the surface of the cell. This could lead to the inhibition of viral internalisation and impairment of the viral replication (Kupferschmidt and Cohen, 2020). This action of NPs is due to their binding to the viral envelope or its protein. This binding prevents the interaction of the virus with the host cell. The efficacy of the treatment is related to the size, shape and the surface charge of the nanoparticles. However, safety measures must be taken regarding the concentration to avoid cytotoxicity of host cells (Singh et al., 2017) (Figure 10.3).

The distribution, metabolism, absorption, excretion of nanoparticles depends upon the hydrophobic/hydrophilic profile of nanoparticles (Patil et al., 2020). The nanoparticles are expelled from the cells mainly by exocytosis. Nanoparticles with diameter of 5nm or less are excreted through urine. The larger nanoparticles are excreted through colon, kidney and liver (Shan et al., 2020).

10.5 APPLICATION OF NANOTHERAPY IN COVID-19 MANAGEMENT

Various mechanisms by which nanoparticles can exert their antiviral effects are depicted in Figure 10.4. These mechanisms could be used to develop new drugs for the management of COVID-19. Antimicrobial drugs tested in clinical trials for COVID-19 include chloroquine, lopinavir, ritonavir, ribavirim and remdesivir (Li et al., 2020) as shown in Table 10.2. Nanoencapsulation of antimicrobial drugs may provide safer treatments for COVID-19 (Mainardes and Diedrich, 2020). Nanostructured systems can be used as adjuvants for vaccines, as well as therapeutics for COVID-19 through the targeting of antiviral drugs (Uskoković V., 2020). Biopolymer nanoparticles like nanocelluloses can be used as therapeutic delivery vehicles. These nano formulations improve stability and provide efficient delivery of drugs to targeted sites and long-term release. The surface of NPs can be modified with different treatments to make them suitable carriers for nano formulations (Patra et al., 2018). Advantages of such NP include controlled therapeutic delivery applications, nontoxicity, biocompatibility and biodegradability. Earlier, nanocelluloses have been used as carriers for various types of antivirals (Gunathilake et al., 2020). Some of the potential applications of nanoparticles in COVID-19 management have shown in Figure 10.4.

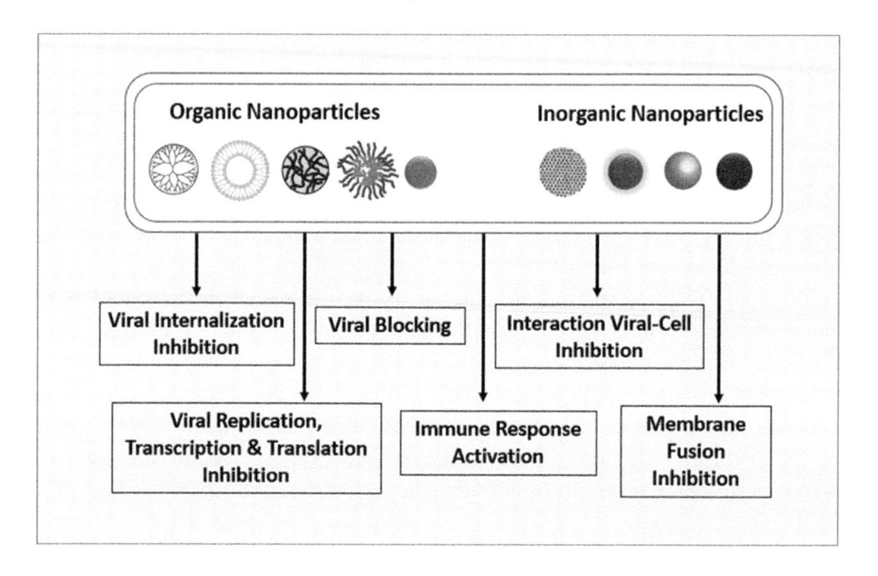

FIGURE 10.4 Applications of nanoparticles in COVID-19 management.

10.6 NANO VACCINE

There are many advantages of nano vaccines. The nano vaccines show immunostimulatory effects, their size and surface properties and drug release can be controlled and they show increased strong humoral and cellular responses (Mamo and Poland, 2012). Owing to their unique nanocarrier characteristics, two lipid mRNA-based COVID-19 vaccines, viz., BNT162b2 and mRNA-1273 have exhibited more than 95% efficacy (Khurana et al., 2021).

10.7 SURFACE DISINFECTION

Coating surfaces with nanomaterials which inactivate viruses can be exploited for manufacture of sterilised equipment. Nanomaterials can be used to manufacture self-disinfecting surfaces and surfaces with inherent antipathogenic properties (Rai et al., 2020; Weiss et al., 2020). Other applications could include manufacture of surfaces which have antiviral, antimicrobial, self-sanitising properties and which cause viral deactivation (Huang et al., 2020). To prevent the spread of COVID-19, surfaces can be coated with nanomaterials like silver, zinc oxide, copper NPs can oxidise and release ions having antimicrobial properties (Weiss et al., 2020)

10.8 PERSONAL PROTECTIVE EQUIPMENT (PPE)

Nanostructures can be used to design efficient and safe PPEs. PPE can be coated with nanomaterials which have self-cleaning, antiviral and antimicrobial

TABLE 10.2
Some Drugs Used in the Treatment of COVID-19

Drug	Characteristics	Limitations
Chloroquine (CQ) and hydroxychloroquine (HCQ) (Anti-malarial drugs)	Anti-inflammatory agents that effectively inhibit the viral entry (Liu et al., 2020)	CQ and HCQ may interfere with the functioning of other drugs or at high doses (Pal et al., 2020; Zou et al., 2020).
	Both are associated with efficient absorption in different tissues (Liu et al., 2020).	HCQ is associated with cardiotoxic effects (e.g., drug-induced cardiomyopathy) and gastrointestinal effects (Pal et al., 2020).
	HCQ is more efficient than CQ, based on molecular docking analysis (Achutha et al., 2020).	
Tocilizumab (TCZ) (IL-6 receptor (IL-6R) antagonist)	TCZ acts by inhibiting IL-6 binding to receptors and decreasing cytokine storm (Zhang et al., 2020). There were no TCZ -linked complications or illness deterioration (Zhang et al., 2020).	TCZ increases the risk of cardiovascular disease (Zhang et al., 2020). TCZ might cause some adverse effects, including hepatic damage, neutropenia, thrombocytopenia or serious secondary infection (Zhang et al., 2020; Kotak et al., 2020).
Remdesivir (RDV) (A prodrug of an adenosine analogue)	RDV is a broad-spectrum antiviral agent and a potent inhibitor of SARS-CoV-2 replication in nasal and bronchial airway epithelial cells (Doggrell, 2020). RDV decreases the recovery time in COVID-19 patients and currently is the first approved anti-viral drug against COVID-19 (Beigel et al., 2020).	RDV might cause some side effects such as the elevation in hepatic enzyme levels and diarrhoea (Grein et al., 2020). The poor oral-bioavailability of RDV hindered its prophylactic use (Malin et al., 2020).
Lopinavir/ritonavir (LPVr) (a protease inhibitor)	LPV/r plays an important role in the "early stage i.e. initial 7–10 days" clinical outcome (Yao et al., 2020). LPV/RTV-treated COVID-19 patients demonstrated a lower mortality rate than HCQ (Karolyi et al., 2021).	LPV might cause some adverse reactions, including elevated hepatic enzymes, increased triglycerides, in addition to diarrhoea, nausea, asthenia and gastrointestinal side effects (Yao et al., 2020). LPVr could increase the risk of bradycardia (Beyls et al., 2020).
Ribavirin	Ribavirin is a broad-spectrum antiviral drug that is available in oral dosage forms (Wang and Zhu, 2020).	Ribavirin, at high doses, increases the risk of hematologic toxicity, besides, its poor in vitro efficacy and poor outcomes. Therefore, ribavirin was not considered a viable treatment for further investigation by the WHO in COVID-19 patients (McCreary and Pogue, 2020).

properties. Nanoparticles and nanofibers with intrinsic hydrophobic water repelling properties and self-cleaning properties can be used to coat PPEs (Rodríguez et al., 2021).

10.9 NANODIAGNOSTICS

Nanotechnology can be used for nucleic acid testing after amplification of nucleic acids with NPs, point-of-care testing through simple colour changes after applying nanostructures (diagnose infected individuals); high sensitivity electrochemical sensors (using NPs); chiral biosensors (conjugating NP with coronavirus specific antibodies), etc (Jindal and Gopinath, 2020; Singh et al., 2020).

Gold NPs when coupled to complementary DNA sequences colour change from red to blue. This is due to formation of a tertiary complex between the viral antigen with the NPs. (Qiu et al., 2020).

Iron oxide NPs can separate the viral RNA from sample solution and can be used to prepare analyte preconcentration, signal amplification and biosensing (Chen et al., 2017). Field-effect transistors based on graphene can detect SARS-CoV-2 within a minute (Seo et al., 2020).

10.10 NANOBIOSENSORS

The SARS-CoV-2 biosensor uses thiol-modified antisense oligonucleotide-capped glyconanoparticles. It can detect COVID-19 positive cases through colour change in less than 10 minutes (Moitra et al., 2020).

10.11 NANOMATERIAL BASED SMART PHONE DETECTORS

Battery-operated and smartphone camera-based amplifications using inorganic quantum dots are poised to become the future SARS-CoV-2 detection systems (Udugama et al., 2020).

10.12 NANOTHERANOSTICS

Theranostics is a field which combines diagnosis with targeted therapy (Filippi et al., 2020). Nuclear imaging using radiolabeled nanomaterials, inorganic and organic NPs nanomaterials and nanosomes are possible candidates for nanotheranostics (Siafaka et al., 2021). Quantum dots in fluorescence imaging helps to visualise cell behaviour and can be used for treatment at the same time (Miyazawa et al., 2021).

10.13 APPLICATIONS USING INORGANIC AND ORGANIC PARTICLES

Organic NPs include liposomes, dendrimers, micelles, etc. Inorganic nanoparticles include transition metal nanoparticles like gold, copper, zinc nanoparticles, metal oxides like iron oxide, zinc oxide nanoparticles and carbon-based nanoparticles. Inhalable organic and inorganic NPs could be potential candidates to target the lung. This would circumvent the side effects from high serum concentrations of conventional administrations (Joseph et al., 2021). Ag, Cu, Zn and metal oxides like Fe_2O_3, TiO_2, ZnO_2 and carbon-based NPs have antipathogenic effects. They interfere with one or more stages of viral life cycle stages (Ruiz-Hitzky et al., 2020). Graphene and its derivatives inactivate the virus by photothermal properties. They bind to the viral S protein, which prevents it from interacting with the host cell receptors (Seifi et al., 2021). Ahmed et al., 2020 have demonstrated that gold NPs enhance the antiviral activity of the drug, which could be used to destroy SARS-CoV-2. Curcumin functionalised gold NPs have demonstrated high antiviral properties for respiratory syncytial virus. These NPs inactivate the virus before it interacts with the cells (Yang et al., 2018). In healthy tissues, the metallic NPs may cause generation of reactive oxygen species, damage to DNA, alter the cellular morphology and cause cytoskeletal defects (Soenen et al., 2011). The conjugation of NPs with polymers or biological components may help to reduce these toxic effects.

10.14 NANO BIOMATERIALS

Biomaterials are materials used in medicine and dentistry that come in contact with living tissue. A common example is tooth filling. Nano-forms of biomaterials can be used to reduce the adverse side effects of conventional therapy. Nano biomaterial therapeutics can be used to deliver drugs directly to the lungs (Cao et al., 2021). Other forms of nano biomaterials like gum-based hydrogels, nanogels, multilayered polyelectrolyte films, DNA aptamers and nanocarriers like nano capsules, nanospheres and polymers are promising candidates to fight against COVID-19 (Vahedifard et al., 2021). Nanobodies can be used to identify pathogens, envelop and neutralise the virus. They can serve as diagnostic or therapeutic tools against the SARS-CoV-2 virus (Adhikari et al., 2020).

10.15 REGULATORY ASPECTS OF NANOPARTICLES

The development of nanostructured systems needs to be regulated and follow regulatory requirements.

Uniform definitions for the identification and application of NPs are required. Also, there is a need for legal provisions concerning NPs and facilitate their marketing. Validated methods for analysis, detection and characterisation are required. There should be complete information about the effects of nanomaterials and nanomaterial exposure (Baran A, 2016). Authorisation of substances and ingredients, qualification of hazardous waste, reinforcing conformity assessment methods, restrictions on the entry of chemical substances, etc., should be part of the nanomaterial's regulation (Baran A, 2016).

10.16 LIMITATIONS

The greatest challenge is transfer of nanomaterial technology to clinical uses and the feasibility of large-scale production (Rai et al., 2020).

Nanomaterials can have a detrimental effect on human health (Singh et al., 2019; Yu et al., 2020). Nanomaterials cause the generation of oxidative stress, reactive nitrogen species, which in turn induce genotoxicity, inflammation, fibrosis, etc. (Kim et al., 2015). Nanomaterials can also cause reactive nitrogen species-mediated damage (Ferreira et al., 2018). Magnetic nanoparticles cause burst drug release and have low stability. This is overcome by attaching surface ligands to magnetic nanoparticles (Yang et al., 2018). Other causes of concern include unpredictable side effects, safety, toxicity, long-term fate and study designs with large sample sizes (Yayehrad et al., 2021). There is insufficient understanding about the cellular, pathogenic and pathophysiologic aspects of SARS-CoV-2 and COVID-19 with the nano bio interfaces involved in drug/vaccine development and delivery (Yayehrad et al., 2021). Another major limitation of using inorganic nanoparticles is that their long-term toxicity and clearance have not been evaluated sufficiently (Núñez et al., 2018).

If adequate rules and legislation are not applied, the use of nanotechnology may cause irreversible damage to the environment and humans (Pandey and Jain, 2020). In spite of their shortcomings, NPs can be used in the management of COVID-19.

10.17 CONCLUSION

Nanotechnology has contributed significantly to overcoming COVID-19. Harnessing the potential of nanotherapeutics holds great promise for better diagnosis, prevention, drug delivery, treatment and control of COVID-19.

REFERENCES

Achutha AS, Pushpa VL, Suchitra S. Theoretical Insights into the Anti-SARS-CoV-2 Activity of Chloroquine and Its Analogs and in Silico Screening of Main Protease Inhibitors. J Proteome Res. 2020 Nov 6;19(11):4706–4717. doi: 10.1021/acs.jproteome.0c00683.

Adhikari SP, Meng S, Wu YJ, Mao YP, Ye RX, Wang QZ, Sun C, Sylvia S, Rozelle S, Raat H, Zhou H. Epidemiology, causes, clinical manifestation and diagnosis, prevention and control of coronavirus disease (COVID-19) during the early outbreak period: A scoping review. Infect Dis Poverty. 2020 Mar 17;9(1):29. doi: 10.1186/s40249-020-00646-x.

Ahmed SF, Quadeer AA, McKay MR. Preliminary Identification of Potential Vaccine Targets for the COVID-19 Coronavirus (SARS-CoV-2) Based on SARS-CoV Immunological Studies. Viruses. 2020 Feb 25;12(3):254. doi: 10.3390/v12030254.

Al-Halifa S, Gauthier L, Arpin D, Bourgault S, Archambault D. Nanoparticle-Based Vaccines Against Respiratory Viruses. Front Immunol. 2019 Jan 24;10:22. doi: 10.3389/fimmu.2019.00022.

Arruda E, Cintra OAL, Hayden FG. Respiratory Tract Viral Infections. Tropical Infectious Diseases. 2006:637–59. doi: 10.1016/B978-0-443-06668-9.50064-8. Epub May 15, 2009. PMCID: PMC7152450.

Baran A. Nanotechnology: Legal and Ethical Issues. Eng Manag Product Serv. 2016;8(1): 47–54. https://doi.org/10.1515/emj-2016-0005.

Bchetnia M, Girard C, Duchaine C, Laprise C. The Outbreak of the Novel Severe Acute Respiratory Syndrome Coronavirus 2 (SARS-CoV-2): A Review of the Current Global Status. J Infect Public Health. 2020 Nov;13(11):1601–1610. doi: 10.1016/j.jiph.2020.07.011.

Beigel JH, Tomashek KM, Dodd LE. Remdesivir for the Treatment of Covid-19—Preliminary Report. Reply. N Engl J Med. 2020 Sep 3;383(10):994. doi: 10.1056/NEJMc2022236.

Beyls C, Martin N, Hermida A, Abou-Arab O, Mahjoub Y. Lopinavir-Ritonavir Treatment for COVID-19 Infection in Intensive Care Unit: Risk of Bradycardia. Circ Arrhythm Electrophysiol. 2020 Aug;13(8):e008798. doi: 10.1161/CIRCEP.120.008798.

Boncristiani HF, Criado MF, Arruda E. Respiratory Viruses. Encyclopedia of Microbiol. 2009:500–18. doi: 10.1016/B978-012373944-5.00314-X. Epub 2009 Feb 17. PMCID: PMC7149556.

Brachman PS. Infectious Diseases-Past, Present, and Future. Int J Epidemiol. 2003 Oct;32(5):684–686. doi: 10.1093/ije/dyg282.

Brian DA, Baric RS. Coronavirus Genome Structure and Replication. Curr Top Microbiol Immunol. 2005;287:1–30. doi: 10.1007/3-540-26765-4_1.

Cao LZ, Wang ZS, Wang B. Application of Nano Biomaterials in Antiviral Vaccine Adjuvant. Chin J Appl Chem. 2021;38(5):572–581.

Chan WCW. Nano Research for COVID-19. ACS Nano. 2020 Apr 28;14(4):3719–3720. doi: 10.1021/acsnano.0c02540.

Chen YT, Kolhatkar AG, Zenasni O, Xu S, Lee TR. Biosensing Using Magnetic Particle Detection Techniques. Sensors (Basel). 2017 Oct 10;17(10):2300. doi: 10.3390/s17102300.

Cheng VC, Lau SK, Woo PC, Yuen KY. Severe Acute Respiratory Syndrome Coronavirus as an Agent of Emerging and Reemerging Infection. Clin Microbiol Rev. 2007 Oct;20(4):660–694. doi: 10.1128/CMR.00023-07.

Coleman CM, Liu YV, Mu H, Taylor JK, Massare M, Flyer DC, Smith GE, Frieman MB. Purified Coronavirus Spike Protein Nanoparticles Induce Coronavirus Neutralizing Antibodies in Mice. Vaccine. May 30, 2014;32(26):3169–3174. doi: 10.1016/j.vaccine.2014.04.016.

Costantino HR, Illum L, Brandt G, Johnson PH, Quay SC. Intranasal Delivery: Physicochemical and Therapeutic Aspects. Int J Pharm. 2007 Jun 7;337(1–2):1–24. doi: 10.1016/j.ijpharm.2007.03.025.

Doggrell SA. Remdesivir, a Remedy or a Ripple in Severe COVID-19? EOID. 2020;29(11):1195–1198. doi: 10.1080/13543784.2020.1821645.

Drosten C, Günther S, Preiser W, van der Werf S, Brodt HR, Becker S, Rabenau H, Panning M, Kolesnikova L, Fouchier RA, Berger A, Burguière AM, Cinatl J, Eickmann M, Escriou N, Grywna K, Kramme S, Manuguerra JC, Müller S, Rickerts V, Stürmer M, Vieth S, Klenk HD, Osterhaus AD, Schmitz H, Doerr HW. Identification of a Novel Coronavirus in Patients with Severe Acute Respiratory Syndrome. N Engl J Med. May 15, 2003;348(20):1967–1976. doi: 10.1056/NEJMoa030747.

Duverger E, Herlem G, Picaud F. A Potential Solution to Avoid Overdose of Mixed Drugs in the Event of Covid-19: Nanomedicine at the Heart of the Covid-19 Pandemic. J Mol Graph Model. 2021 May;104:107834. doi: 10.1016/j.jmgm.2021.107834.

Ferreira CA, Ni D, Rosenkrans ZT, Cai W. Scavenging of Reactive Oxygen and Nitrogen Species with Nanomaterials. Nano Res. 2018 Oct;11(10):4955–4984. doi: 10.1007/s12274-018-2092-y.

Filippi L, Chiaravalloti A, Schillaci O, Cianni R, Bagni O. Theranostic approaches in nuclear medicine: Current status and future prospects. Expert Rev Med Devices. 2020 Apr;17(4):331–343.

Gautret P, Lagier JC, Parola P, Hoang VT, Meddeb L, Mailhe M, Doudier B, Courjon J, Giordanengo V, Vieira VE, Tissot Dupont H, Honoré S, Colson P, Chabrière E, La Scola B, Rolain JM, Brouqui P, Raoult D. Hydroxychloroquine and Azithromycin as a Treatment of COVID-19: Results of an Open-Label Non-Randomized Clinical Trial. Int J Antimicrob Agents. 2020 Jul;56(1):105949. doi: 10.1016/j.ijantimicag.2020.105949.

Gavriatopoulou M, Korompoki E, Fotiou D, Ntanasis-Stathopoulos I, Psaltopoulou T, Kastritis E, Terpos E, Dimopoulos MA. Organ-Specific Manifestations of COVID-19 Infection. Clin Exp Med. 2020 Nov;20(4):493–506. doi: 10.1007/s10238-020-00648-x.

Glebov OO. Understanding SARS-CoV-2 endocytosis for COVID-19 drug repurposing. FEBS J. 2020 Sep;287(17):3664–3671. doi: 10.1111/febs.15369.

Gregory AE, Titball R, Williamson D. Vaccine Delivery Using Nanoparticles. Front Cell Infect Microbiol. 2013 Mar 25;3:13. doi: 10.3389/fcimb.2013.00013.

Grein J, Ohmagari N, Shin D, Diaz G, et al. Compassionate Use of Remdesivir for Patients with Severe Covid-19. N Eng J Med. 2020;382(24):2327–2336. doi: 10.1056/NEJMoa2007016.

Gunathilake TMSU, Ching YC, Chuah CH, Abd Rahman N, Nai-Shang L. Recent Advances in Celluloses and Their Hybrids for Stimuli-Responsive Drug Delivery. Int J Biol Macromol. 2020;158:670–688.

Gunathilake TMSU, Ching YC, Uyama H, Chuah CH. Nanotherapeutics for Treating Coronavirus Diseases. J Drug Deliv Sci Technol. 2021 Aug;64:102634. doi: 10.1016/j.jddst.2021.102634.

Huang H, Fan C, Li M, Nie HL, Wang FB, Wang H, Wang R, Xia J, Zheng X, Zuo X, Huang J. COVID-19: A Call for Physical Scientists and Engineers. ACS Nano. 2020 Apr 28;14(4):3747–3754. doi: 10.1021/acsnano.0c02618.

Jindal S, Gopinath P. Nanotechnology Based Approaches for Combatting COVID-19 Viral Infection. Nano Express. 2020;1(2):13.

Joseph SKMAA, Thomas S, Nair SC. State of the Art Nanotechnology Based Drug Delivery Strategies to Combat COVID-19. Int J Appl Pharm. 2021;13(3) :18–29. https://doi.org/10.22159/ijap.2021v13i3.40865.

Karolyi M, Pawelka E, Mader T, Omid S, Kelani H, Ely S, Jilma B, Baumgartner S, Laferl H, Ott C, Traugott M, Turner M, Seitz T, Wenisch C, Zoufaly A. Hydroxychloroquine Versus Lopinavir/Ritonavir in Severe COVID-19 Patients: Results from a Real-Life Patient Cohort. Wien Klin Wochenschr. 2021 Apr;133(7–8):284–291. doi: 10.1007/s00508-020-01720-y.

Khan M, Kazmi S, Bashir A, Siddique N. COVID-19 infection: Origin, transmission, and characteristics of human coronaviruses. J Adv Res. 2020;24:91–98.

Khurana A, Allawadhi P, Khurana I, Allwadhi S, Weiskirchen R, Banothu AK, Chhabra D, Joshi K, Bharani KK. Role of Nanotechnology Behind the Success of mRNA Vaccines for COVID-19. Nano Today. 2021 Jun;38:101142. doi: 10.1016/j.nantod.2021.101142.

Kim KS, Lee D, Song CG, Kang PM. Reactive Oxygen Species-Activated Nanomaterials as Theranostic Agents. Nanomedicine (Lond). 2015;10(17):2709–2723. doi: 10.2217/nnm.15.108.

Kotak S, Khatri M, Malik M, Malik M, Hassan W, Amjad A, Malik F, Hassan H, Ahmed J, Zafar M. Use of Tocilizumab in COVID-19: A Systematic Review and Meta-Analysis of Current Evidence. Cureus. 2020 Oct 9;12(10):E10869. doi: 10.7759/cureus.10869.

Kupferschmidt K, Cohen J. Race to Find COVID-19 Treatments Accelerates. Science. 2020 Mar 27;367(6485):1412–1413. doi: 10.1126/science.367.6485.1412.

Li H, Liu SM, Yu XH, Tang SL, Tang CK. Coronavirus Disease 2019 (COVID-19): Current Status and Future Perspectives. Int J Antimicrob Agents. 2020 May;55(5):105951. doi: 10.1016/j.ijantimicag.2020.105951.

Liu J, Cao R, Xu M, Wang X, Zhang H, Hu H, Li Y, Hu Z, Zhong W, Wang M. Hydroxychloroquine, a Less Toxic Derivative of Chloroquine, Is Effective in Inhibiting SARS-CoV-2 Infection in Vitro. Cell Discov. 2020 Mar 18;6:16. doi: 10.1038/s41421-020-0156-0.

Lu R, Zhao X, Li J, Niu P, Yang B, Wu H, Wang W, Song H, Huang B, Zhu N, Bi Y, Ma X, Zhan F, Wang L, Hu T, Zhou H, Hu Z, Zhou W, Zhao L, Chen J, Meng Y, Wang J, Lin Y, Yuan J, Xie Z, Ma J, Liu WJ, Wang D, Xu W, Holmes EC, Gao GF, Wu G, Chen W, Shi W, Tan W. Genomic Characterisation and Epidemiology of 2019 Novel Coronavirus: Implications for Virus Origins and Receptor Binding. Lancet. 2020 Feb 22;395(10224):565–574. doi: 10.1016/S0140-6736(20)30251-8.

Lukassen S, Chua RL, Trefzer T, Kahn NC, Schneider MA, Muley T, Winter H, Meister M, Veith C, Boots AW, Hennig BP, Kreuter M, Conrad C, Eils R. SARS-CoV-2 Receptor ACE2 and TMPRSS2 Are Primarily Expressed in Bronchial Transient Secretory Cells. EMBO J. May 18, 2020;39(10):e105114. doi: 10.15252/embj.20105114.

Mainardes RM, Diedrich C. The Potential Role of Nanomedicine on COVID-19 Therapeutics. Ther Deliv. 2020 Jul;11(7):411–414. doi: 10.4155/tde-2020-0069.

Malin JJ, Suárez I, Priesner V, Fätkenheuer G, Rybniker J. Remdesivir Against COVID-19 and Other Viral Diseases. Clin Microbiol Rev. 2020 Oct 14;34(1):e00162–E00220. doi: 10.1128/CMR.00162-20.

Mamo T, Poland GA. Nanovaccinology: The Next Generation of Vaccines [Me]ets 21st Century Materials Science and Engineering. Vaccine. 2012 Oct 19;30(47):6609–6611. doi: 10.1016/j.vaccine.2012.08.023.

McCreary EK, Pogue JM. Coronavirus Disease 2019 Treatment: A Review of Early and Emerging Options. Open Forum Infect Dis. 2020 Mar 23;7(4):ofaa105. doi: 10.1093/ofid/ofaa105.

McKay PF, Hu K, Blakney AK, Samnuan K, Brown JC, Penn R, Zhou J, Bouton CR, Rogers P, Polra K, Lin PJC, Barbosa C, Tam YK, Barclay WS, Shattock RJ. Self-Amplifying RNA SARS-CoV-2 Lipid Nanoparticle Vaccine Candidate Induces High Neutralizing Antibody Titers in Mice. Nat Commun. 2020 Jul 9;11(1):3523. doi: 10.1038/s41467-020-17409-9.

Melchor-Martínez EM, Torres Castillo NE, Macias-Garbett R, Lucero-Saucedo SL, Parra-Saldívar R, Sosa-Hernández JE. Modern World Applications for Nano-Bio Materials: Tissue Engineering and COVID-19. Front Bioeng Biotechnol. May 14, 2021;9:597958. doi: 10.3389/fbioe.2021.597958.

Miyazawa T, Itaya M, Burdeos GC, Nakagawa K, Miyazawa T. A Critical Review of the Use of Surfactant-Coated Nanoparticles in Nanomedicine and Food Nanotechnology. Int J Nanomedicine. 2021 Jun 9;16:3937–3999. doi: 10.2147/IJN.S298606.

Moitra P, Alafeef M, Dighe K, Frieman MB, Pan D. Selective Naked-Eye Detection of SARS-CoV-2 Mediated by N Gene Targeted Antisense Oligonucleotide Capped Plasmonic Nanoparticles. ACS Nano. 2020 Jun 23;14(6):7617–7627. doi: 10.1021/acsnano.0c03822.

Morens DM, Folkers GK, Fauci AS. The Challenge of Emerging and Re-Emerging Infectious Diseases. Nature. 2004 Jul 8;430(6996):242–249. doi: 10.1038/nature02759. Erratum in: Nature. 2010 Jan 7;463(7277):122.

Núñez C, Estévez SV, Del Pilar Chantada M. Inorganic Nanoparticles in Diagnosis and Treatment of Breast Cancer. J Biol Inorg Chem. 2018 May;23(3):331–345. doi: 10.1007/s00775-018-1542-z.

Pal M, Berhanu G, Desalegn C, Kandi V. Severe Acute Respiratory Syndrome Coronavirus-2 (SARS-CoV-2): An Update. Cureus. 2020 Mar 26;12(3):e7423. doi: 10.7759/cureus.7423.

Pandey G, Jain P. Assessing the Nanotechnology on the Grounds of Costs, Benefits, and Risks. Beni-Suef Univ J Basic Appl Sci. 202 0;9:63. https://doi.org/10.1186/s43088-020-00085-5.

Patil VM, Singhal S, Masand N. A Systematic Review on Use of Aminoquinolines for the Therapeutic Management of COVID-19: Efficacy, Safety and Clinical Trials. Life Sci. 2020 Aug 1;254:117775. doi: 10.1016/j.lfs.2020.117775. Epub May 11, 2020. PMID: 32418894; PMCID: PMC7211740.

Patra JK, Das G, Fraceto LF, Campos EVR, Rodriguez-Torres MDP, Acosta-Torres LS, Diaz-Torres LA, Grillo R, Swamy MK, Sharma S, Habtemariam S, Shin HS. Nano Based Drug Delivery Systems: Recent Developments and Future Prospects. J Nanobiotechnol. 2018 Sep 19;16(1):71. doi: 10.1186/s12951-018-0392-8.

Petrosillo N, Viceconte G, Ergonul O, Ippolito G, Petersen E. COVID-19, SARS and MERS: Are They Closely Related? Clin Microbiol Infect. 2020 Jun;26(6):729–734. doi: 10.1016/j.cmi.2020.03.026.

Pimentel TA, Yan Z, Jeffers SA, Holmes KV, Hodges RS, Burkhard P. Peptide Nanoparticles as Novel Immunogens: Design and Analysis of a Prototypic Severe Acute Respiratory Syndrome Vaccine. Chem Biol Drug Des. 2009 Jan;73(1):53–61. doi: 10.1111/j.1747-0285.2008.00746.x.

Qiu G, Gai Z, Tao Y, Schmitt J, Kullak-Ublick GA, Wang J. Dual-Functional Plasmonic Photothermal Biosensors for Highly Accurate Severe Acute Respiratory Syndrome Coronavirus 2 Detection. ACS Nano. May 26, 2020;14(5):5268 5277. doi: 10.1021/acsnano.0c02439.

Raghuwanshi D, Mishra V, Das D, Kaur K, Suresh MR. Dendritic Cell Targeted Chitosan Nanoparticles for Nasal DNA Immunization Against SARS CoV Nucleocapsid Protein. Mol Pharm. 2012 Apr 2;9(4):946–956. doi: 10.1021/mp200553x.

Rai M, Bonde S, Yadav A, Plekhanova Y, Reshetilov A, Gupta I, Golińska P, Pandit R, Ingle AP. Nanotechnology-Based Promising Strategies for the Management of COVID-19: Current Development and Constraints. Expert Rev Anti Infect Ther. 2020 Nov 8:1–10. doi: 10.1080/14787210.2021.1836961.

Rodríguez NB, Formentini G, Favi C, Marconi M. Environmental Implication of Personal Protection Equipment in the Pandemic Era: LCA Comparison of Face Masks Typologies. Procedia CIRP. 2021;98:306–311. doi: 10.1016/j.procir.2021.01.108.

Rothan HA, Byrareddy SN. The Epidemiology and Pathogenesis of Coronavirus Disease (COVID-19) Outbreak. J Autoimmun. 2020 May;109:102433. doi: 10.1016/j.jaut.2020.102433.

Ruiz-Hitzky E, Darder M, Wicklein B, Ruiz-Garcia C, Martín-Sampedro R, Del Real G, Aranda P. Nanotechnology Responses to COVID-19. Adv Healthc Mater. 2020 Oct;9(19):e2000979. doi: 10.1002/adhm.202000979.

Seah I, Agrawal R. Can the Coronavirus Disease 2019 (COVID-19) Affect the Eyes? A Review of Coronaviruses and Ocular Implications in Humans and Animals. Ocul Immunol Inflamm. 2020 Apr 2;28(3):391–395. doi: 10.1080/09273948.2020.1738501.

Seifi T, Reza Kamali A. Antiviral Performance of Graphene-Based Materials with Emphasis on COVID-19: A Review. Med Drug Discov. May 25, 2021:100099. doi: 10.1016/j.medidd.2021.100099.

Sekimukai H, Iwata-Yoshikawa N, Fukushi S, Tani H, Kataoka M, Suzuki T, Hasegawa H, Niikura K, Arai K, Nagata N. Gold Nanoparticle-Adjuvanted S Protein Induces a Strong Antigen-Specific IgG Response Against Severe Acute Respiratory Syndrome-Related Coronavirus Infection, but Fails to Induce Protective Antibodies and Limit Eosinophilic Infiltration in Lungs. Microbiol Immunol. 2020 Jan;64(1): 33–51. doi: 10.1111/1348-0421.12754.

Seo G, Lee G, Kim MJ, Baek SH, Choi M, Ku KB, Lee CS, Jun S, Park D, Kim HG, Kim SJ, Lee JO, Kim BT, Park EC, Kim SI. Rapid Detection of COVID-19 Causative Virus (SARS-CoV-2) in Human Nasopharyngeal Swab Specimens Using Field-Effect Transistor-Based Biosensor. ACS Nano. 2020 Apr 28;14(4):5135–5142. doi: 10.1021/acsnano.0c02823.

Shan C, Yao YF, Yang XL, Zhou YW, Gao G, Peng Y, Yang L, Hu X, Xiong J, Jiang RD, Zhang HJ, Gao XX, Peng C, Min J, Chen Y, Si HR, Wu J, Zhou P, Wang YY, Wei HP, Pang W, Hu ZF, Lv LB, Zheng YT, Shi ZL, Yuan ZM. Infection with Novel Coronavirus (SARS-CoV-2) Causes Pneumonia in Rhesus Macaques. Cell Res. 2020 Aug;30(8):670–677. doi: 10.1038/s41422-020-0364-z.

Shim BS, Park SM, Quan JS, Jere D, Chu H, Song MK, Kim DW, Jang YS, Yang MS, Han SH, Park YH, Cho CS, Yun CH. Intranasal Immunization with Plasmid DNA Encoding Spike Protein of SARS-Coronavirus/Polyethylenimine Nanoparticles Elicits Antigen-Specific Humoral and Cellular Immune Responses. BMC Immunol. 2010 Dec 31;11:65. doi: 10.1186/1471-2172-11-65.

Siafaka PI, Okur NÜ, Karantas ID, Okur ME, Gündoğdu EA. Current Update on Nanoplatforms as Therapeutic and Diagnostic Tools: A Review for the Materials Used as Nanotheranostics and Imaging Modalities. Asian J Pharm Sci. 2021 Jan;16(1):24–46. doi: 10.1016/j.ajps.2020.03.003.

Singh AV, Kumar M, Dubey AK. Effect of Pre-Existing Diseases on COVID-19 Infection and Role of New Sensors and Biomaterials for Its Detection and Treatment. Med Devices Sens. 2020 Oct 28:E10140. doi: 10.1002/mds3.10140.

Singh AV, Laux P, Luch A, Sudrik C, Wiehr S, Wild AM, Santomauro G, Bill J, Sitti M. Review of Emerging Concepts in Nanotoxicology: Opportunities and Challenges for Safer Nanomaterial Design. Toxicol Mech Methods. 2019 Jun;29(5):378–387. doi: 10.1080/15376516.2019.1566425.

Singh L, Kruger HG, Maguire GEM, Govender T, Parboosing R. The Role of Nanotechnology in the Treatment of Viral Infections. Ther Adv Infect Dis. 2017 Jul;4(4):105–131. doi: 10.1177/2049936117713593.

Soenen S, Rivera-Gil P, Montenegro JM, Parak WJ, De Smedt S, Braeckmans K. Cellular Toxicity of Inorganic Nanoparticles: Common Aspects and Guidelines for Improved Nanotoxicity Evaluation. Nano Today. 2011;6(5):4 46–465. https://doi.org/10.1016/j.nantod.2011.08.001.

Udugama B, Kadhiresan P, Kozlowski HN, Malekjahani A, Osborne M, Li VYC, Chen H, Mubareka S, Gubbay JB, Chan WCW. Diagnosing COVID-19: The Disease and Tools for Detection. ACS Nano. 2020 Apr 28;14(4):3822–3835. doi: 10.1021/acsnano.0c02624.

Ullah MA, Araf Y, Sarkar B, Moin AT, Reshad RA, Hasanur MD. Pathogenesis, Diagnosis and Possible Therapeutic Options for COVID-19. J Clin Exp Invest. 2020;11(4):em00755. doi:10.29333/jcei/8564.

Usković V. Why Have Nanotechnologies Been Underutilized in the Global Uprising Against the Coronavirus Pandemic? Nanomedicine (Lond). 2020 Jul;15(17):1719–1734. doi: 10.2217/nnm-2020-0163.

Vahedifard F, Chakravarthy K. Nanomedicine for COVID-19: The Role of Nanotechnology in the Treatment and Diagnosis of COVID-19. Emergent Mater. 2021;4(1):75–99. doi: 10.1007/s42247-021-00168-8.

Wang Y, Zhu LQ. Pharmaceutical Care Recommendations for Antiviral Treatments in Children with Coronavirus Disease 2019. World J Pediatr. 2020 Jun;16(3):271–274. doi: 10.1007/s12519-020-00353-5.

Weiss C, Carriere M, Fusco L, Capua I, Regla-Nava JA, Pasquali M, Scott JA, Vitale F, Unal MA, Mattevi C, Bedognetti D, Merkoçi A, Tasciotti E, Yilmazer A, Gogotsi Y, Stellacci F, Delogu LG. Toward Nanotechnology-Enabled Approaches Against the COVID-19 Pandemic. ACS Nano. 2020 Jun 23;14(6):6383–6406. doi: 10.1021/acsnano.0c03697.

Yang D. Application of Nanotechnology in the COVID-19 Pandemic. Int J Nanomedicine. 2021 Jan 26;16:623–649. doi: 10.2147/IJN.S296383.

Yang HY, Li Y, Lee DS. Multifunctional and Stimuli-Responsive Magnetic Nanoparticle-Based Delivery Systems for Biomedical Applications. AdvTher. 2018;1(2):1800011.

Yao TT, Qian JD, Zhu WY, Wang Y, Wang GQ. A Systematic Review of Lopinavir Therapy for SARS Coronavirus and MERS Coronavirus-A Possible Reference for Coronavirus Disease-19 Treatment Option. J Med Virol. 2020 Jun;92(6):556–563. doi: 10.1002/jmv.25729.

Yayehrad AT, Siraj EA, Wondie GB, Alemie AA, Derseh MT, Ambaye AS. Could Nanotechnology Help to End the Fight Against COVID-19? Review of Current Findings, Challenges and Future Perspectives. Int J Nanomedicine. 2021 Aug 24;16:5713–5743. doi: 10.2147/IJN.S327334.

Yu Z, Li Q, Wang J, Yu Y, Wang Y, Zhou Q, Li P. Reactive Oxygen Species-Related Nanoparticle Toxicity in the Biomedical Field. Nanoscale Res Lett. May 20, 2020;15(1):115. doi: 10.1186/s11671-020-03344-7.

Zhang C, Wu Z, Li JW, Zhao H, Wang GQ. Cytokine Release Syndrome in Severe COVID-19: Interleukin-6 Receptor Antagonist Tocilizumab May Be the Key to Reduce Mortality. Int J Antimicrob Agents. 2020 May;55(5):105954. doi: 10.1016/j.ijantimicag.2020.105954.

Zou L, Dai L, Zhang X, Zhang Z, Zhang Z. Hydroxychloroquine and Chloroquine: A Potential and Controversial Treatment for COVID-19. Arch Pharm Res. 2020 Aug;43(8):765–772. doi: 10.1007/s12272-020-01258-7.

11 Pharmacogenomics and COVID-19

Rishabh Chaudhary, Divya Chaudhary and Vikas Mishra

11.1 INTRODUCTION

The coronavirus disease 2019 (COVID-19) epidemic, which was caused by a novel coronavirus known as SARS-CoV-2 (severe acute respiratory syndrome coronavirus 2), was first found in Wuhan, Hubei Province, People's Republic of China, in December 2019 and has since spread to many nations. The COVID-19 outbreak is a global emergency since it has been shown to have significant morbidity and mortality, a high contagious propensity, rapid growth rate and a broad range of severity. (1) The causative agent of COVID-19, SARS-CoV-2, is a member of the B lineage of the *Betacoronavirus* genera of *Coronaviridae* family. SARS-CoV-2 is a highly transmissible, enveloped, positive-sense single-stranded RNA virus of about 30 kb in length. (2) Vascular endothelial cells, airway and alveolar epithelial cells and alveolar macrophages are one of the potential targets of SARS-CoV-2 virus once it reaches the respiratory system. The clinical signs of COVID-19 disease range from asymptomatic or moderate illness to severe, extreme and even fatal life-threatening flu-like symptoms which may progress to pneumonia, acute respiratory distress, renal failure and death, particularly in older (≥60 years) individuals. (3,4) To lessen the devastating effects of the COVID-19 outbreak on patients, a large number of therapeutic drugs are explored thoroughly and taken under investigational trials without well-established effectiveness or safety data evidence. Moreover, there is also scant information on the pharmacogenomics of these drugs, regardless of the fact that genetic variables play a very important role in the effectiveness and/or toxicity of drugs. (5,6) In numerous research contexts, the hypothesis that genetic variability leads to heterogeneity in disease symptoms and medication responses is universally recognised and proven. For certain medications, genetic makeup has significant implications for pharmacological treatment to reduce toxicity and maximise responsiveness. (7,8) The field of pharmacogenomics may enable the customisation of therapeutic agents, as it primarily sheds light on the identification and authentication of gene variants that directly or indirectly influence the effects of drug usually via modifications in pharmacokinetic (drug absorption, distribution, metabolism and excretion) as well as the pharmacodynamic profile through modifying the drug's target or by reconfiguring the cellular mechanisms which eventually determine a person's responsivity to the drug's pharmacological effects (Figure 11.1). Thus,

DOI: 10.1201/9781003324911-16

enhancing the effectiveness and safety of drugs as well as decreasing the risk for unexpected toxicities. (9) In severely ill patients who cannot afford a failure of a poor treatment strategy, pharmacogenomics may assist health care professionals in selecting appropriate and suitable first-line medications and since COVID-19 is more likely to cause damage to patients who are already receiving treatment for other conditions, it is essential that the associated risks of drug toxicity be kept to

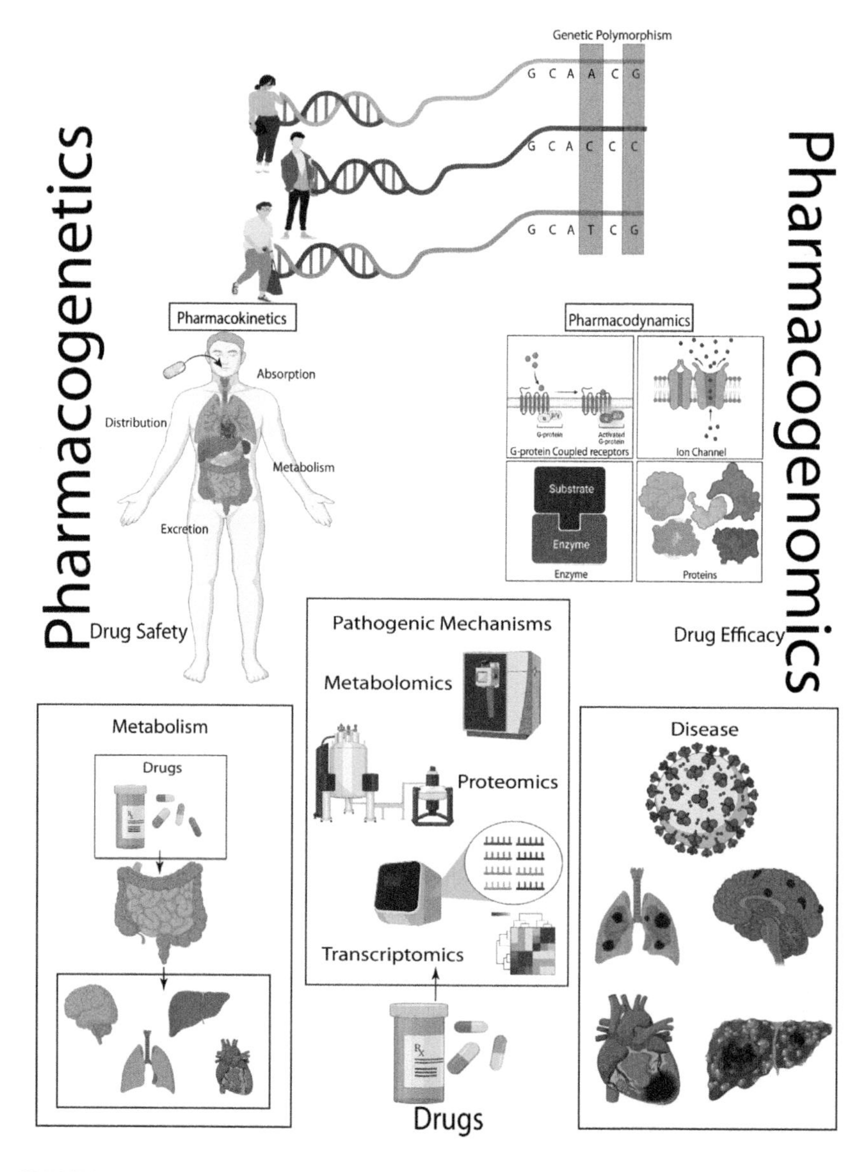

FIGURE 11.1 A schematic overview of pharmacogenomics.

a minimum. (10) This chapter provides a brief overview of the pharmacogenomic literature as well as the clinical recommendations that are currently available for COVID-19 candidate drug therapies.

11.2 A DETAILED OVERVIEW OF PHARMACOGENOMICS (PGx)

In principle, pharmacogenomics is the study of gene-related variations in drug-metabolising enzymes, drug transporters, receptors, targets as well as how these variations in genes produce drug-related phenotypes, including drug response and drug toxicity. (11) The revolution in the field of genomics led to a significant increase in our understanding of molecular structure, function, evolution, mapping of genomes as well as the development of techniques for efficiently collecting massive amounts of genetic data. (12) Similarly, advances in therapeutics have led to the development of drug therapy that may effectively treat or manage a variety of diseases ranging from small bacterial infection to cancer. (13) Nonetheless, the emergence of such effective and potent drug candidates has also raised the significance of inter-individual variability in therapeutic responses, which may extend from life-threatening adverse reactions to a failure of targeted clinical effect on the other side of the continuum. Simultaneously, the use of conventional genomic tools led to the discovery that genetic inheritance/heredity is an essential component in individual variance in therapeutic response to drugs. (14) The primary aim of pharmacogenomics is to use information about a patient's genetic code (nucleotide sequence) to improve drug treatment modalities in order to increase efficacy and efficiency, target medications solely to those individuals who are likely to respond and reduce drug-related adverse events. (15) Moreover, pharmacogenomics helps to figure out how variations in genes affect therapeutic effectiveness and toxicity profile. Such type of investigational research may elucidate how variability in genetic makeup influences the pharmacokinetic and pharmacodynamic profile of a particular drug.

In the late 1950s, when an inherited glucose-6-phosphate dehydrogenase (G6PD) deficiency was found to cause severe haemolysis in some patients exposed to the primaquine (anti-malarial drug), the belief that drug response is determined by genetic factors that ultimately modify the pharmacokinetic and pharmacodynamic properties of the drug was first conceived. (16) The term *'pharmacogenetics'* was coined by German geneticist Friedrich Vogel in 1959, and it was supported by groundbreaking research of Elliott S. Vesell and John G. Page, proving that the pharmacokinetic character profile of antipyrine drug is more similar in monozygotic twins relative to dizygotic twins. (17) Multiple family studies undertaken between the 1960s and 1980s showed inheritance patterns for a number of pharmacological effects, eventually leading to molecular research that discovered the hereditary factors for a number of characteristics. (18–20) In 1987, the first polymorphic drug-metabolising gene, *Cytochrome P450 2D6 (CYP2D6)*, was cloned and characterised. In the 1990s, the promising therapeutic benefit of pharmacogenomics was successfully shown for a number of genes. (21,22) Inherited deficiency of the *TPMT (Thiopurine Methyltransferase)* gene, which encodes the enzyme thiopurine S-methyltransferase, in many people have been found to exhibit haematopoietic toxicity upon administration of mercaptopurine and azathioprine (antileukemic and immunosuppressive) drugs, though enactment of this discovery in the hospitals progressed imperceptibly at that time. (23) The Human Genome Project (1990 to 2003) and state of art technology for genome-wide variation analysis have accelerated the pace of pharmacogenetic discoveries. These discoveries enhanced the research process and facilitated agonistic genome-wide sequence analyses in patients with distinct drug characteristics, frequently resulting in the detection of novel genetic variations significantly related with drug therapeutic effects. These advanced genomic technologies also aided in the introduction of the word 'pharmacogenomics' into the pharmacology vocabulary. (24) Nonetheless, all these discoveries resulting from genomic techniques require proper validation before being translated for clinical use. The process of evaluation and validation can be aided by understanding the molecular factors that govern how genetic variance tends to affect drug therapeutic responses. Genes usually tend to vary according to ethnic background, lineage and ancestry, limiting the clinical translation of pharmacogenetic features from one group to the other. The inter-individual dosing variability of anticoagulant drug warfarin, for example, are significantly influenced by genetic variations in the *VKORC1 (Vitamin K epOxide Reductase Complex)* and *CYP2C9 (Cytochrome P450 2C9)* genes. (25) Also, it is now quite evident that many pharmacological properties are strongly controlled by dozens of different variations in the very same gene, some of which are infrequent, as well as variations in different genes within the same patient. Pharmacogenomics research at the National Institutes of Health (NIH) in the United States and the 100,000 Genomes Project (www.1000Genomes.org) in England are two of many continuing initiatives to make genome discoveries and translation into treatments relatively easy. Their results might later be utilised to improve pharmaceutical list and dosage for individual patients. The identification and study of inherited therapeutic response factors and somatically acquired genomic variations are integral pharmacogenomic elements in various programmes. (9)

11.3 GENE POLYMORPHISM AND PHARMACOGENOMIC RESEARCH

The gene polymorphism constitutes the structural framework of pharmacogenetic research which refers to genetic variations that influence how a patient responds to treatment. (26) Our knowledge of the genetic basis of individual therapeutic responses has progressed, owing to extensive studies on genes and therapies associated to an array of diseases. Alterations in the DNA sequence within the genetic material that affects a single nucleotide to megabases are referred to as genetic variations. Single nucleotide polymorphisms (SNPs), gene deletions and insertions (indels) and copy number variations (CNVs) are all some of the most common genetic variants that have been studied. These genetic variants occur naturally in the genome and are the result of defects or errors in DNA replication process during cell division. However, exogenous agents like chemical mutagens or viruses may also cause alterations in the genetic code. SNPs are the most common type of inherited genome variation (constituting ~90% of all human genetic variation) among individuals. As a result of these genetic variances, different clinical outcomes might occur among monogenic and complex disease patients who are prescribed the same drug. There are different types of polymorphism occuring in the genes, which encodes phase 1 CYP450 (cytochrome P450) or phase 2 drug-metabolising enzymes (for example, glutathione S-transferases, UDP-glucuronosyltransferases, sulfotransferases, N-acetyltransferases and methyltransferases) drug targets, drug transporters or human leucocyte antigen (HLA) alleles and predict toxicity or efficacy of drug molecules. (27,28) Genetic variations of *CYP2D6*, *CYP2C9*, *CYP3A5* and *CYP2C19* have been linked with the therapeutic effects of tamoxifen, phenytoin, tacrolimus and clopidogrel, respectively. (29–32) The dosage regimen of chemotherapeutic drug irinotecan and mercaptopurines should be adjusted according to the polymorphism of *UGT1A1* (UDP glucuronosyltransferase 1A1) and *TPMT* gene, respectively. (33,34) Small candidate-gene analyses of genes encoding drug-metabolising enzymes led to early advancements in the pharmacogenomics field. Several investigations explored the significance of a single SNP or a group of SNPs in one or more candidate genes in a number of healthcare settings. Previously, genetic variations in thiopurine methyltransferase (*TPMT*) gene and its influence on thiopurine treatment for acute lymphoblastic leukaemia and immunological regulation is one of the best-known and well-established genotypic and phenotypic relationship. Numerous research demonstrated that the *TPMT* genotype or phenotype may be used to label those patients who are at high risk of haematological toxicity following thiopurine treatment, which has spurred the speed of clinical research in TPMT pharmacogenomics. (35,36) The hepatic cytochrome P450 gene *CYP2D6*, which catalyses the metabolic process of numerous medications, is another potential gene. The analgesic compound codeine, a prodrug which is bioactivated by CYP2D6 to morphine, a potent opioid agonist, is one drug whose metabolism is tightly bound to CYP2D6 genotype or phenotype. The efficacy and safety profile of codeine has been shown to be strongly influenced by CYP2D6 polymorphisms. (37,38) In addition, alterations in drug-metabolising genes are not the only vital factor that can influence pharmacological efficacy and efficiency; genetic polymorphisms in genes that code for drug receptors, drug transporters and drug targets could have the same negative influence. For instance, a common promoter mutation in *VKORC1* gene, a molecular target of anticoagulant drug warfarin, has a substantial impact on the doses needed by individuals. *VKORC1* gene encodes vitamin K epoxide reductase, the enzyme that primarily inhibits warfarin activities. There are significant associations that exist between *VKORC1* gene variants and warfarin sensitivity and lower dosage demands. (39) Similarly, it was shown that a number of drug transporters also have pharmacogenomic interactions with the pharmacokinetics or effects of drug molecules. For instance, a synonymous

SNP (a coding SNP which does not alter the amino acid sequence) in the *ABCB1* (ATP-binding cassette sub-family B member 1) gene has been related with the significantly increased concentration of digoxin. (40) Similarly, genetic polymorphism in the influx transporter *SLCO1B1* (Solute carrier organic anion transporter family member 1B1) have been linked to a variety of characteristics, including a high risk of statin-induced myopathy, methotrexate-related adverse drug reactions like gastrointestinal toxicity and disposition of flavopiridol (a cyclin-dependent kinase inhibitor). (41–43) A downside of the conventional candidate-gene methodology is its dependence on pre-existing and previous data of the pharmacology of the drug and the pathophysiology of the illness or phenotypic traits. This may lead to a hindrance, with discoveries primarily restricted to only a handful of selected genes implicated in certain genomic features of significance (44). As a result, advanced state-of-the-art technologies like whole-genome sequencing or genome-wide association studies (GWAS) are highly sophisticated approaches for pharmacogenomic research of complex diseases and medications, where the detailed genomic structure underlying their aetiology is unknown and the genetic causes are highly polygenic.

Recent breakthroughs in sequencing technologies and GWAS approach have resulted in the introduction of a multitude of gene datasets, markedly redefining the outlook of pharmacogenomic research. Organisations like The Clinical Pharmacogenetics Implementation Consortium (CPIC) and The Pharmacogenomics Knowledgebase (PharmGKB), which employ standardised ways to analyse the research process and give clinical information, are critical for pharmacogenetics to become part of everyday clinical practise and patient management. pharmgkb.org and cpicpgx.org are two important online information websites that give clinicians concise therapeutic-related information and recommendations as well as pharmacogenomic associations. Several other essential resources include the Pharmacogenomics And Cell database (PACdb) (www.PACdb.org/) and SNP and CNV Annotation (SCAN) database (www.scandb.org), which were established and presented by the PGx of Anticancer Agents Research Group. PACdb is a cell line dataset (lymphoblastoid cell lines; LCLs) collecting pharmacology associated databases, including genetic expression, genotypes and pharmacological data. (45) SCAN database is created to store, collect, annotate and demonstrate the connection link between gene and gene expression profiles. Additionally, SCAN is a highly extensive genomic and genetic dataset encompassing CNV and SNP annotations. (46) In the past few years, scientific evidence has significantly strengthened up, several new guidelines have emerged, and genomic treatments are now gradually integrating into everyday healthcare system for various disease conditions.

11.4 PHARMACOGENOMICS OF ANTI-COVID-19 DRUGS

11.4.1 Chloroquine and Hydroxychloroquine

The aminoquinoline category drugs chloroquine and hydroxychloroquine were the first anti-viral therapeutic agents officially authorised for COVID-19 disease on March 28, 2020. (47) These two medications are at the frontline of COVID-19 therapeutic choices according to early positive data exhibiting antiviral action against SARS-CoV-2 at micromolar doses in tissue culture as well as their clinical benefit found in observational studies of a small number of patients. It has been reported that the effective concentration (EC50) of chloroquine and hydroxychloroquine against SARS-CoV-2 in Vero E6 cell line is 2.71 M and 4.51 M, respectively. (48) However, on June 15, 2020, it was withdrawn from the market owing to updated evidence and large observational clinical trials that suggested the drug's potential complications. (49,50) Chloroquine phosphate and hydroxychloroquine sulphate have been prescribed to treat amoebic liver abscess, malaria infection as well as various autoimmune diseases for many years. It is believed to raise the phagolysosome's pH, which hinders the process of virus fusion and stops the virus from attaching to the receptors. (50) It is possible that the regulatory actions of these immunologic agents will be of assistance in the management of the cytokine storm driven by COVID-19 pandemic. The US Food and Drug Administration (FDA) has highlighted concerns over these therapeutic agents' potential to cause cardiac abnormalities when used to treat COVID-19 condition. (51) According to the findings of a large retrospective study conducted on COVID-19 patients who were hospitalised in New York, it was found that patients who were treated with the drug combination consisting of hydroxychloroquine and azithromycin had a significantly higher risk of cardiac arrest as well as an increased death rate. (52) Currently, the National Institutes of Health (NIH) recommendations do not suggest using chloroquine or hydroxychloroquine treatments in COVID-19. Chloroquine and hydroxychloroquine accumulate in the autophagosomes and lysosomes of phagocytic cells, affecting the pH concentration locally, inhibiting the expression of the Interleukin 1, major histocompatibility complex class II (MHC-II), tumour necrosis factor, interferons as well as interfering with synthase activity of cyclic GMP-AMP pathway. (53) In the metabolism of chloroquine and hydroxychloroquine, the cytochrome P450 (CYP) enzymes, particularly CYP3A4, CYP2C8 and CYP2D6 play an important role. (54) Both chloroquine and hydroxychloroquine go via the CYP isozyme-mediated process of N-dealkylation. (55) Solute carrier organic anion (*SLCO*) gene is responsible for encoding the organic anion transporting polypeptides (OATP), which are influx cellular membrane transporters. Both chloroquine and hydroxychloroquine are substrates for OATP. According to PharmGKB database, the genes that have a critical role in the pharmacokinetics of the drugs chloroquine and hydroxychloroquine are very important pharmacogenes (VIPs). In the field of pharmacogenomics, these genes have been identified as being of the utmost significance.

According to a study by Lee et al., the percentage of the active metabolite of hydroxychloroquine to its parent medication rose by 20% in individuals with systemic lupus erythematosus condition who carried variations in the cytochrome P450 2D6 enzyme (rs1135840: CC versus GG was 0.90 versus 0.69, respectively, $p < 0.01$). (56) Another study demonstrated that low-activity of CYP2C8 alleles (CYP2C8*2 and CYP2C8*3 and CYP2C8*4) were linked to a poorer reduction in gametocytemia than homozygous wild-type allele CYP2C8*1A one day following primaquine/chloroquine treatment in a cohort of 164 malaria patients (-2.21 versus -11.18 gametocytes/L, p = 0.007, respectively). (57) Additionally, in the very same cohort group, a second study revealed that, after accounting for CYP2C8, variant alleles of *SLCO1A2* and *SLCO1B1* exhibited poorer gametocytemia clearance than the wild-type alleles (p = 0.018 and 0.024, respectively). (58) Hepatic cytochrome P450 CYP2D6, whose expression varies between populations as a result of genetic polymorphisms, is responsible for the metabolism of both hydroxychloroquine and chloroquine. Individuals with weak or intermediate CYP2D6 metabolising abilities (CYP2D6*4; CYP2D6*10) have been shown to have excessive hydroxychloroquine levels in their bodies. (56) They may have the potential to enhance the activity of other CYP2D6-related substrates, like carvedilol and labetalol, while simultaneously lowering the efficacy of prodrugs that depend on CYP2D6 enzyme for their stimulation, for example, tramadol and codeine. (59,60)

11.4.2 Lopinavir and Ritonavir

Lopinavir, chemically known as [N-(4(S)-(2-(2,6-dimethylphenoxy)-acetylamino)-3(S)-hydroxy-5-phenyl-1(S)-benzylpentyl)-3-methyl-2(S)-(2-oxo (1,3diazaperhydroinyl) butanamin)], is a HIV-1 protease (HIV-1 PR) inhibitor, which prevents the cleavage of a polyprotein named Gag-Pol (HIV-1 M:B_HXB2R), resulting in the production of immature, non-infectious virus particles. (61) Lopinavir is typically used to treat human immunodeficiency virus (HIV), but because it also prevents a post-entry stage in the MERS-CoV (Middle East respiratory syndrome–related coronavirus) replication cycle, it has potential as a COVID-19 therapy. Both of these viruses have been successfully curbed by lopinavir treatment. (62) However, both HIV-1 PR inhibitors (lopinavir and ritonavir) are not currently recommended for COVID-19 treatment owing to a lack of safety and effectiveness in a small randomised controlled trial as well as a problem of poor drug exposure in comparison to what is needed for the inhibition of coronavirus (SARS-CoV-2). (63) Ritonavir inhibits the CYP3A-mediated metabolism of lopinavir, which results in a prolonged exposure to the viral particles. CYP3A is the primary enzyme responsible for lopinavir metabolic degradation, whereas ABCB1 and ABCC2 (efflux transporters) are accountable for its cellular transport. The enzymes CYP2D6, CYP3A4, CYP3A5 and CYP2J2 are responsible for the metabolism of ritonavir. Ritonavir has the ability to trigger a number of different CYP isoforms while simultaneously inhibiting CYP3A pathway as well as OATP1B1 and OATPB3 (organic anion transporting polypeptides). (64) Ritonavir strengthened the efficacy and potency of the drugs that are typically metabolised by CYP3A4 pathway, including lopinavir, as well as the concentrations of those drugs. The biological transformation of certain compounds that are metabolised by the glucuronidation process that is mediated by UDP-glucuronosyltransferases (UGT) enzyme is accelerated by ritonavir as well. Thus, it is necessary to conduct research to evaluate the therapeutic significance of such outcomes. (65) Increased CYP3A4 activity is associated with the CYP3A4 polymorphism L292P (rs28371759, CYP*18B), which facilitates the metabolism of lopinavir. L292P was found to be more common in Eastern Asian populations, suggesting that they could metabolise lopinavir and ritonavir more quickly. The *ABCB1* gene has approximately 66 coding SNPs (single nucleotide polymorphisms). (66) The amino acid isoleucine is unlikely to be affected by the 3435 C > T transition, which is located in exon 26 of the *ABCB1* gene and involves a C > T transition. Comparing the Asian populace to the Caucasian and African populace, the variant allele frequency of 3435 C > T is notably different. (67) The *ABCB1* gene contains a large number of non-synonymous polymorphisms. For example, S893A (rs2032582), S400N (rs2229109) and N21D (rs9282564) have the potential to raise the concentration of the drug by lowering ABCB1's efflux activity. Carriers of the *S893A* gene may be found in all groups at a higher frequency, with the percentage reaching as high as 90% in African population. Carriers with the S893T subtype are mainly from eastern region of Asia, whereas carriers with the N21D subtype are from European continent. It is possible that these people will react better to medications that are delivered through ABCB1. (68) Moreover, in order to fully comprehend the pharmacokinetics of lopinavir, it is essential to take into account not only CYP3A pathway but also other different CYP-related enzymes and drug transporters. A pharmacogenomic investigation of lopinavir and ritonavir, which included the examination of 1,380 different variations in 638 HIV-positive Caucasian individuals, uncovered four important variants. In a population pharmacokinetic-pharmacogenetic model, the clearance/removal of lopinavir was greater in persons who had SLCO1B1*4/*4 and lower in those who had two or more variant alleles of SLCO1B1*5 (rs4149056), ABCC2 (rs717620) or a CYP3A (rs6945984) tag in comparison to the reference group (12.6 versus 3.9 versus 5.4 l/h, respectively, $p < 0.01$). (69) A whole other genome-wide association study investigated 290 variants for their effects on lopinavir/ritonavir combination-related toxicity effects in 104 HIV-infected Caucasian patients. Genetic variations in the *CETP* (Cholesteryl ester transfer protein), *MCP1* (Monocyte chemoattractant protein-1), *ABCC2* (ATP-binding cassette subfamily C member 2), *LEP* (leptin), and *SLCO1B3* genes were associated with hyperbilirubinemia and dyslipidemia and variations in the interleukin-6 gene was associated with diarrhoea (all $p < 0.01$). (70)

11.4.3 Azithromycin

Azithromycin, an azalide, is a member of macrolide family of antibiotics with a broad gram -ve and gram +ve spectrum. It is primarily used to treat a wide variety of skin and respiratory infections with immunoregulatory and anti-inflammatory effects, causing the regulation of both adaptive and innate immune system activities. (71) A recent study by Gautret et al. was demonstrated that the use of azithromycin plus hydroxychloroquine combination on respiratory viral loads in COVID-19 patients is potentially more efficacious and shows a synergistic effect relative to hydroxychloroquine treatment alone. (72) Despite the fact that macrolide substances (for example, erythromycin and clarithromycin) are associated with a large number of drug-drug related interactions, azithromycin has fewer interactions. Studies reported that azithromycin is not a major substrate of SLCO1B1/SLCO1B3, CYP3A4, and as a result, it has very fewer drug-drug related interactions. (73) However, Zing and colleagues done research that demonstrated the impact of P-glycoprotein transporter encoded by the *ABCB1* (ATP-binding cassette subfamily B member 1) gene on the pharmacokinetics of azithromycin in the Chinese Han population. (74) After a single dose administration of azithromycin, genetic variation in *ABCB1* gene demonstrated up to a two-fold lower concentration in twenty healthy individuals (rs2032582TT/rs1045642TT versus rs2032582GG/rs1045642CC was 468.0 versus 911.2 ng/ml, respectively, $p = 0.013$). (74) According to the results, different genetic variations in the *ABCB1* gene are certainly a significant factor in the inter-individual variability in the pharmacodynamics and pharmacokinetics of azithromycin. Additionally, due to the additive effects that chloroquine and hydroxychloroquine compound have on prolonged QT interval, an increased amount of azithromycin's systemic exposure is of considerable importance when it is taken with either chloroquine or hydroxychloroquine. This combination can produce catastrophic arrhythmic events.

11.4.4 Corticosteroids

Corticosteroids, such as dexamethasone and hydrocortisone are now recommended for patients with severe COVID-19, including those with COVID-related acute respiratory distress syndrome (ARDS). Since new clinical evidence reveal that corticosteroids enhance survival rate in severe COVID-19, there has been a re-emergence of investigation in the effects of corticosteroids typically in the treatment of ARDS. In addition, the results of recent randomised controlled studies in seriously ill patients with COVID-19 demonstrate that the administration of systemic corticosteroids is associated with a lower risk of passing away (odds ratio [OR] 0.66, 95% confidence interval [CI] 0.53–0.82). (75) However, in spite of the fact that a number of randomised controlled studies have demonstrated the beneficial effects of corticosteroids in the treatment of ARDS, practise varies. A number of different genetic variants were linked with corticosteroids response and toxicities in numerous diseases, as well as genes which are involved in the binding of receptors (such as *NR3C1* (Nuclear Receptor Subfamily 3 Group C Member 1) and *CRHR1* (Corticotropin-releasing hormone receptor 1), chaperone/co-chaperone protein (like FKBP5 (FK506 binding protein 5), STIP1 (Stress Induced Phosphoprotein 1), ST13 (suppression of tumorigenicity 13), drug transporters (such as ABCB1 and MDR1 (multidrug resistance protein 1) and metabolic enzymes (for example, GSTT1 (glutathione S-transferase theta 1), CYP3A5, CYP3A7, CYP3A4). (76) The molecular mechanistic and metabolic signalling of steroidal drugs are intricately complex and diverse, as well as genetic determinants with enough clinical evidence for medical use to COVID-19 are not yet well-known. Genetic variants related with corticosteroids in PharmGKB that had a degree of evidence greater than level 3 (low) were evaluated. These genetic variants were only evaluated when they were used in combination treatment or when they were inhaled. It was unable to locate any pharmacogenetic research that focused explicitly on the efficacy of corticosteroids in ARDS treatment. (77)

11.4.5 Renin–Angiotensin–Aldosterone System (RAAS) Inhibitors

The discovery that the severely ill COVID-19 individuals are much more likely to have a history of severe cardiac-related pathologies, diabetes and/or hypertension and take RAAS inhibitors, which prompted the theory that imbalance of RAAS mechanism may have an important role. There are various pharmacogenetic determinants of RAAS inhibitor disposition and response. Thus, it is possible to analyse the disposition and responsiveness of renin-angiotensin inhibiting drugs using pharmacogenetics. (78) CYP2C9 is responsible for the metabolism of the ACE receptor blocker named losartan. It is possible that the genotyping of some CYP2C9 polymorphisms would greatly impact the therapeutic value of losartan. CYP2C9*2 (Arg144Cys) and CYP2C9*3 (Ile359Leu) are the genetic variations with the most frequent lowered levels of function. Additionally, it is possible that knowing the genotype of CYP2C9 can help find the optimal dose of the drug. Other pharmacogenetic indicators, including polymorphisms of the *ABCB1* gene, have the potential to be helpful predictors of the response to losartan. (79) The three most significant *ABCB1* gene polymorphisms are G2677T/A (rs2032582), C1236T (rs1128502) and C3435T (rs1045642). (80) Moreover, regarding the angiotensin-converting enzyme (ACE) inhibitors, ACE rs1799752 is associated with variation in ACE inhibitors. (81) Genetic polymorphism in *ABCB1* gene consists of a fifty-nucleotide long deletion. There was a correlation between those with deletion/deletion diplotype and a poorer clinical result. The considerable pharmacogenetic predictor for the efficacy of spironolactone is rs4961, which is situated in the *ADD1* (α-adducin) gene and predicts medication response when combined with furosemide. (82) In comparison to those who had the T allele, those who carried the G allele responded to treatment more effectively. (83)

11.4.6 Janus Kinase (JAK) Inhibitors

An important immunomodulator drug for COVID-19 treatment that is now being tested in a number of clinical trials is one that inhibits non-receptor tyrosine kinases named Janus kinases (JAKs). JAK inhibitors, in comparison to other candidate molecules for the management of COVID-19, have a significant pharmacological attribute for a potentially viable drug repurposing: easy drug administration, a beneficial pharmacokinetic profile, as well as versatile pharmacodynamic properties by exerting dual anti-inflammatory and anti-viral effects. These pharmacological characteristics are necessary for an effective drug repurposing. (84) Orally selective Janus-associated kinase 1 and 2 (JAK1 and 2) inhibitor ruxolitinib has been authorised by the FDA in 2014 for the treatment of myelofibrosis as well as graft versus host disease (GvHD), an immune-mediated disease and baricitinib has been recommended for rheumatoid arthritis (RA) treatment. (85) There has been no research published that studied the influence of different genetic variations on either of the two medications in any patient population. Nevertheless, the pharmacokinetic mechanism of these drugs involves few pharmacogenes that may be considered significant, for example, CYP3A4. Ruxolitinib is a prominent substrate of CYP3A4, while baricitinib is minor. (77) CYP2C9 is also involved in the partial metabolism of ruxolitinib. Each of these genes are regarded as very important pharmacogenes in the PharmGKB, and particularly notable genetic variations can be found in both of them. (77) However, in spite of the fact that the primary pharmacogene of baricitinib, SLC22A8, which encodes the OAT3 (organic anion transporter 3) transporter, is not a very important pharmacogene, the impact of genetic variability on its cellular and molecular activity has been described in another substrate molecule. (86)

11.5 CONCLUDING REMARKS

The field of pharmacogenetic and pharmacogenomics possess the potential to contribute to the implementation of the goal of tailored medical treatment in a number of critical diseases that are both effective and less harmful. Moreover, the addition of proteomics and metabolomics into this area of study, as well as advancements in the fields of pharmacogenetics and pharmacogenomics, will undoubtedly lead to more effective outcomes. Individual variation in drug dose response is caused by a number of different factors apart from genetic inheritance; however, significant advances in the field of molecular genetics and pharmacological research have highlighted the potential of offering the health professional with credible information that could make it possible to design drug selection and/or dose regime to the likely response of the patient to that specific class of drug or that dose based on the patient's genotype. Nevertheless, despite the ardour positivity that has surrounded pharmacogenomics and pharmacogenetics research, the use of these two fields in clinical settings has progressed at a very gradual pace. In the meantime, with the aim of lowering adverse drug effects and improving patient's better health outcomes, the next step of pharmacogenomic investigation might very well educate scientists about the complex genetic design of variable drug response and might even possibly discover genes and molecular as well as cellular pathways which could indeed be used as therapeutic targets for newer drugs.

11.6 CONFLICT OF INTEREST

The authors declare no conflict of interest.

REFERENCES

1. Yang L, Liu S, Liu J, Zhang Z, Wan X, Huang B, et al. COVID-19: Immunopathogenesis and Immunotherapeutics. Signal Transduct Target Ther. 2020 Jul;5(1):128.
2. Harrison AG, Lin T, Wang P. Mechanisms of SARS-CoV-2 Transmission and Pathogenesis. Trends Immunol. 2020 Dec;41(12):1100–1115.
3. Wang D, Hu B, Hu C, Zhu F, Liu X, Zhang J, et al. Clinical Characteristics of 138 Hospitalized Patients with 2019 Novel Coronavirus-Infected Pneumonia in Wuhan, China. JAMA—J Am Med Assoc. 2020 Mar;323(11):1061–1069.
4. Chan JFW, Yuan S, Kok KH, To KKW, Chu H, Yang J, et al. A Familial Cluster of Pneumonia Associated with the 2019 Novel Coronavirus Indicating Person-to-Person Transmission: A Study of a Family Cluster. Lancet. 2020;395(10223):514–523.
5. Baden LR, Rubin EJ. Covid-19—The Search for Effective Therapy. N Eng J Med. 2020;382:1851–1852.
6. Wu R, Wang L, Kuo HCD, Shannar A, Peter R, Chou PJ, et al. An Update on Current Therapeutic Drugs Treating COVID-19. Curr Pharmacol Reports. 2020;6(3):56–70.
7. Weinshilboum R. Inheritance and Drug Response. N Engl J Med. 2003 Feb;348(6):529–537.
8. Evans WE, McLeod HL. Pharmacogenomics—Drug Disposition, Drug Targets, and Side Effects. N Engl J Med. 2003 Feb;348(6):538–549.
9. Relling MV, Evans WE. Pharmacogenomics in the Clinic. Nature. 2015 Oct;526(7573):343–350.
10. Yang X, Yu Y, Xu J, Shu H, Xia J, Liu H, et al. Clinical Course and Outcomes of Critically Ill Patients with SARS-CoV-2 Pneumonia in Wuhan, China: A single-Centered, Retrospective, Observational Study. Lancet Respir Med. 2020 May;8(5):475–481.
11. Crews KR, Hicks JK, Pui CH, Relling MV, Evans WE. Pharmacogenomics and Individualized Medicine: Translating Science into Practice. Clin Pharmacol Ther. 2012 Oct;92(4):467–475.
12. Lander ES, Linton LM, Birren B, Nusbaum C, Zody MC, Baldwin J, et al. Initial Sequencing and Analysis of the Human Genome. Nature. 2001 Feb;409(6822):860–921.
13. Weinshilboum RM. The Therapeutic Revolution. Clin Pharmacol Ther [Internet]. 1987;42(5):481–484. http://europepmc.org/abstract/MED/3315388.
14. Walker NF. Pharmacogenetics Heredity and the Response to Drugs. Can J Genet Cytol. 1964;6(1):118–119.
15. Weinshilboum R, Wang L. Pharmacogenomics: Bench to Bedside. Nat Rev Drug Discov. 2004 Sep;3(9):739–748.
16. Beutler E. Drug-Induced Hemolytic Anemia. Pharmacol Rev. 1969 Mar;21(1):73–103.
17. Vesell ES, Page JG. Genetic Control of Drug Levels in Man: Antipyrine. Science. 1968 Jul;161(3836):72–73.
18. Price Evans DA, Manley KA, McKusick VA. Genetic Control of Isoniazid Metabolism in Man. Br Med J. 1960 Aug;2(5197):485–491.
19. Blum M, Demierre A, Grant DM, Heim M, Meyer UA. Molecular Mechanism of Slow Acetylation of Drugs and Carcinogens in Humans. Proc Natl Acad Sci U S A. 1991 Jun;88(12):5237–5241.
20. Vatsis KP, Martell KJ, Weber WW. Diverse Point Mutations in the Human Gene for Polymorphic N-Acetyltransferase. Proc Natl Acad Sci U S A. 1991 Jul;88(14):6333–6337.
21. Ingelman-Sundberg M. Pharmacogenomic Biomarkers for Prediction of Severe Adverse Drug Reactions. N Eng J Med U S. 2008;358:637–639.

22. Wang L, McLeod HL, Weinshilboum RM. Genomics and Drug Response. N Engl J Med. 2011 Mar;364(12):1144–1153.
23. Marshall E. Preventing Toxicity With a Gene Test. Science [Internet]. 2003;302 (5645):588–590. www.science.org/doi/abs/10.1126/science.302.5645.588.
24. Carr DF, Alfirevic A, Pirmohamed M. Pharmacogenomics: Current State-of-the-Art. Genes (Basel). 2014 May;5(2):430–443.
25. Pirmohamed M, Kamali F, Daly AK, Wadelius M. Oral Anticoagulation: A Critique of Recent Advances and Controversies. Trends Pharmacol Sci. 2015 Mar;36(3):153–163.
26. Patil J. Pharmacogenetics and Pharmacogenomics: A Brief Introduction. J Pharmacovigil. 2015;3(3).
27. Tomalik-Scharte D, Lazar A, Fuhr U, Kirchheiner J. The Clinical Role of Genetic Polymorphisms in Drug-Metabolizing Enzymes. Pharmacogenomics J. 2008 Feb;8(1):4–15.
28. Lee Ventola C. Role of Pharmacogenomic Biomarkers in Predicting and Improving Drug Response: Part 1: The Clinical Significance of Pharmacogenetic Variants. P T. 2013 Sep;38(9):545–560.
29. Wigle TJ, Jansen LE, Teft WA, Kim RB. Pharmacogenomics Guided-Personalization of Warfarin and Tamoxifen. J Pers Med [Internet]. 2017;7(4). www.mdpi.com/2075-4426/7/4/20.
30. Silvado CE, Terra VC, Twardowschy CA. CYP2C9 Polymorphisms in Epilepsy: Influence on Phenytoin Treatment. Pharmgenomics Pers Med. 2018;11:51–58.
31. Yu M, Liu M, Zhang W, Ming Y. Pharmacokinetics, Pharmacodynamics and Pharmacogenetics of Tacrolimus in Kidney Transplantation. Curr Drug Metab. 2018;19(6):513–522.
32. Zhang YJ, Li MP, Tang J, Chen XP. Pharmacokinetic and Pharmacodynamic Responses to Clopidogrel: Evidences and Perspectives. Int J Environ Res Public Health. 2017 Mar;14(3).
33. Fujita K, Sparreboom A. Pharmacogenetics of Irinotecan Disposition and Toxicity: A Review. Curr Clin Pharmacol. 2010 Aug;5(3):209–217.
34. Relling MV., Klein TE, Gammal RS, Whirl-Carrillo M, Hoffman JM, Caudle KE. The Clinical Pharmacogenetics Implementation Consortium: 10 Years Later. Clin Pharmacol Ther. 2020 Jan;107(1):171–175.
35. Yates CR, Krynetski EY, Loennechen T, Fessing MY, Tai HL, Pui CH, et al. Molecular Diagnosis of Thiopurine S-Methyltransferase Deficiency: Genetic Basis for Azathioprine and Mercaptopurine Intolerance. Ann Intern Med [Internet]. 1997 Apr;126(8):608–614. https://doi.org/10.7326/0003-4819-126-8-199704150-00003.
36. Lennard L, Lilleyman JS, Van Loon J, Weinshilboum RM. Genetic Variation in Response to 6-Mercaptopurine for Childhood Acute Lymphoblastic Leukaemia. Lancet [Internet]. 1990;336(8709):225–229. www.sciencedirect.com/science/article/pii/014067369091745V.
37. Kirchheiner J, Schmidt H, Tzvetkov M, Keulen JT, Lötsch J, Roots I, et al. Pharmacokinetics of Codeine and Its Metabolite Morphine in Ultra-Rapid Metabolizers Due to CYP2D6 Duplication. Pharmacogenomics J. 2007 Aug;7(4):257–265.
38. Lötsch J, Rohrbacher M, Schmidt H, Doehring A, Brockmöller J, Geisslinger G. Can Extremely Low or High Morphine Formation from Codeine Be Predicted Prior to Therapy Initiation? Pain. 2009 Jul;144(1–2):119–124.
39. Rieder MJ, Reiner AP, Gage BF, Nickerson DA, Eby CS, McLeod HL, et al. Effect of VKORC1 Haplotypes on Transcriptional Regulation and Warfarin Dose. N Engl J Med. 2005 Jun;352(22):2285–2293.

40. Hoffmeyer S, Burk O, Von Richter O, Arnold HP, Brockmöller J, Johne A, et al. Functional Polymorphisms of the Human Multidrug-Resistance Gene: Multiple Sequence Variations and Correlation of One Allele with P-Glycoprotein Expression and Activity in Vivo. Proc Natl Acad Sci U S A. 2000 Mar;97(7):3473–3478.
41. Link E, Parish S, Armitage J, Bowman L, Heath S, Matsuda F, et al. SLCO1B1 Variants and Statin-Induced Myopathy—A Genomewide Study. N Engl J Med. 2008 Aug;359(8):789–799.
42. Radtke S, Zolk O, Renner B, Paulides M, Zimmermann M, Möricke A, et al. Germline Genetic Variations in Methotrexate Candidate Genes Are Associated with Pharmacokinetics, Toxicity, and Outcome in Childhood Acute Lymphoblastic Leukemia. Blood. 2013 Jun;121(26):5145–5153.
43. Ni W, Ji J, Dai Z, Papp A, Johnson AJ, Ahn S, et al. Flavopiridol Pharmacogenetics: Clinical and Functional Evidence for the Role of SLCO1B1/OATP1B1 in Flavopiridol Disposition. PLoS One. 2010 Nov;5(11):e13792.
44. Zhu M, Zhao S. Candidate Gene Identification Approach: Progress and Challenges. Int J Biol Sci. 2007 Oct;3(7):420–427.
45. Gamazon ER, Duan S, Zhang W, Huang RS, Kistner EO, Dolan ME, et al. PACdb: A Database for Cell-Based Pharmacogenomics. Pharmacogenet Genomics. 2010 Apr;20(4):269–273.
46. Gamazon ER, Huang RS, Cox NJ. SCAN: A Systems Biology Approach to Pharmacogenomic Discovery. Methods Mol Biol. 2013;1015:213–224.
47. Karalis V, Ismailos G, Karatza E. Chloroquine Dosage Regimens in Patients with COVID-19: Safety Risks and Optimization Using Simulations. Saf Sci. 2020 Sep;129:104842.
48. Yao X, Ye F, Zhang M, Cui C, Huang B, Niu P, et al. In Vitro Antiviral Activity and Projection of Optimized Dosing Design of Hydroxychloroquine for the Treatment of Severe Acute Respiratory Syndrome Coronavirus 2 (SARS-CoV-2). Clin Infect Dis. 2020 Jul;71(15):732–739.
49. Martinez MA. Clinical Trials of Repurposed Antivirals for SARS-CoV-2. Antimicrob Agents Chemother. 2020 Aug;64(9).
50. Magagnoli J, Narendran S, Pereira F, Cummings T, Hardin JW, Sutton SS, et al. Outcomes of Hydroxychloroquine Usage in United States Veterans Hospitalized with Covid-19. medRxiv : The Preprint Server for Health Sciences. 2020.
51. Bajpai J, Pradhan A, Singh A, Kant S. Hydroxychloroquine and COVID-19—A Narrative Review. Indian J Tuberc. 2020 Dec;67(4):S147–S154.
52. Rosenberg ES, Dufort EM, Udo T, Wilberschied LA, Kumar J, Tesoriero J, et al. Association of Treatment with Hydroxychloroquine or Azithromycin with In-Hospital Mortality in Patients with COVID-19 in New York State. JAMA—J Am Med Assoc. 2020 Jun;323(24):2493–2502.
53. Schrezenmeier E, Dörner T. Mechanisms of Action of Hydroxychloroquine and Chloroquine: Implications for Rheumatology. Nat Rev Rheumatol. 2020 Mar;16(3):155–166.
54. Elewa H, Wilby KJ. A Review of Pharmacogenetics of Antimalarials and Associated Clinical Implications. Eur J Drug Metab Pharmacokinet. 2017 Oct;42(5):745–756.
55. Lim HS, Im JS, Cho JY, Bae KS, Klein TA, Yeom JS, et al. Pharmacokinetics of Hydroxychloroquine and Its Clinical Implications in Chemoprophylaxis Against Malaria Caused by Plasmodium Vivax. Antimicrob Agents Chemother. 2009 Apr;53(4):1468–1475.
56. Lee JY, Vinayagamoorthy N, Han K, Kwok SK, Ju JH, Park KS, et al. Association of Polymorphisms of Cytochrome P450 2D6 with Blood Hydroxychloroquine Levels in Patients with Systemic Lupus Erythematosus. Arthritis Rheumatol. 2016 Jan;68(1):184–190.

57. Sortica VA, Lindenau JD, Cunha MG, Ohnishi M Do, Ventura AMR, Ribeiro-Dos-Santos ÂK, et al. The Effect of SNPs in CYP450 in Chloroquine/Primaquine Plasmodium Vivax Malaria Treatment. Pharmacogenomics. 2016 Nov;17(17):1903–1911.
58. Sortica VA, Lindenau JD, Cunha MG, O Ohnishi MD, R Ventura AM, Ribeiro-Dos-Santos ÂKC, et al. SLCO1A2, SLCO1B1 and SLCO2B1 polymorphisms influences chloroquine and primaquine treatment in Plasmodium vivax malaria. Pharmacogenomics. 2017 Oct;18(15):1401–11.
59. Somer M, Kallio J, Pesonen U, Pyykkö K, Huupponen R, Scheinin M. Influence of Hydroxychloroquine on the Bioavailability of Oral Metoprolol. Br J Clin Pharmacol. 2000 Jun;49(6):549–554.
60. Kirchheiner J, Keulen JTHA, Bauer S, Roots I, Brockmöller J. Effects of the CYP2D6 Gene Duplication on the Pharmacokinetics and Pharmacodynamics of Tramadol. J Clin Psychopharmacol [Internet]. 2008;28(1):78–83. https://journals.lww.com/psychopharmacology/Fulltext/2008/02000/Effects_of_the_CYP2D6_Gene_Duplication_on_the.14.aspx.
61. Paintsil E, Cheng YC. Antiviral Agents. Encyclopedia of Microbiology. 2009. pp. 223–257.
62. Costanzo M, De Giglio MAR, Roviello GN. SARS-CoV-2: Recent Reports on Antiviral Therapies Based on Lopinavir/Ritonavir, Darunavir/Umifenovir, Hydroxychloroquine, Remdesivir, Favipiravir and other Drugs for the Treatment of the New Coronavirus. Curr Med Chem. 2020;27(27):4536–4541.
63. Bethesda MD. Coronavirus Disease 2019 (COVID-19) Treatment Guidelines. Bethesda. 2021.
64. Larson KB, Wang K, Delille C, Otofokun I, Acosta EP. Pharmacokinetic Enhancers in HIV Therapeutics. Clin Pharmacokinet. 2014 Oct;53(10):865–872.
65. Young BE, Ong SWX, Kalimuddin S, Low JG, Tan SY, Loh J, et al. Epidemiologic Features and Clinical Course of Patients Infected with SARS-CoV-2 in Singapore. JAMA—J Am Med Assoc. 2020 Apr;323(15):1488–1494.
66. Wolf SJ, Bachtiar M, Wang J, Sim TS, Chong SS, Lee CGL. An Update on ABCB1 Pharmacogenetics: Insights from a 3D Model into the Location and Evolutionary Conservation of Residues Corresponding to SNPs Associated with Drug Pharmacokinetics. Pharmacogenomics J [Internet]. 2011;11(5):315–325. https://doi.org/10.1038/tpj.2011.16.
67. Balram C, Sharma A, Sivathasan C, Lee EJD. Frequency of C3435T Single Nucleotide MDR1 Genetic Polymorphism in an Asian Population: Phenotypic-Genotypic Correlates. Br J Clin Pharmacol. 2003 Jul;56(1):78–83.
68. Tulsyan S, Mittal RD, Mittal B. The Effect of ABCB1 Polymorphisms on the Outcome of Breast Cancer Treatment. Pharmgenomics Pers Med. 2016;9:47–58.
69. Lubomirov R, Lulio J Di, Fayet A, Colombo S, Martinez R, Marzolini C, et al. ADME Pharmacogenetics: Investigation of the Pharmacokinetics of the Antiretroviral Agent Lopinavir Coformuiated with Ritonavir. Pharmacogenet Genomics [Internet]. 2010 Apr;20(4):217–230. https://doi.org/10.1097/FPC.0b013e328336eee4.
70. Aspiroz EL, Figueroa SEC, Cruz R, Porras Hurtado GL, Martín AF, Hurlé ADG, et al. Toxicogenetics of Lopinavir/Ritonavir in HIV-INFECTED EUROPEAN PATIENTS. Per Med. 2014 May;11(3):263–272.
71. Beigel JH, Tomashek KM, Dodd LE, Mehta AK, Zingman BS, Kalil AC, et al. Remdesivir for the Treatment of Covid-19—Final Report. N Engl J Med. 2020 Nov;383(19):1813–1826.
72. Gautret P, Lagier JC, Parola P, Hoang VT, Meddeb L, Mailhe M, et al. Hydroxychloroquine and Azithromycin as a Treatment of COVID-19: Results of an open-label non-randomized clinical trial. Int J Antimicrob Agents. 2020 Jul;56(1):105949.

73. Fohner AE, Sparreboom A, Altman RB, Klein TE. PharmGKB Summary: Macrolide Antibiotic Pathway, Pharmacokinetics/Pharmacodynamics. Pharmacogenet Genomics. 2017 Apr;27(4):164–167.
74. He XJ, Zhao LM, Qiu F, Sun YX, Li-Ling J. Influence of ABCB1 Gene Polymorphisms on the Pharmacokinetics of Azithromycin Among Healthy Chinese Han Ethnic Subjects. Pharmacol Reports. 2009;61(5):843–850.
75. Sterne JAC, Murthy S, Diaz JV, Slutsky AS, Villar J, Angus DC, et al. Association Between Administration of Systemic Corticosteroids and Mortality Among Critically Ill Patients with COVID-19: A Meta-Analysis. JAMA—J Am Med Assoc. 2020 Oct;324(13):1330–1341.
76. Song QQ, Xie WY, Tang YJ, Zhang J, Liu J. Genetic Variation in the Glucocorticoid Pathway Involved in Interindividual Differences in the Glucocorticoid Treatment. Pharmacogenomics. 2017 Feb;18(3):293–316.
77. Takahashi T, Luzum JA, Nicol MR, Jacobson PA. Pharmacogenomics of COVID-19 therapies. NPJ Genomic Med. 2020;5(1):35.
78. Gemmati D, Tisato V. Genetic Hypothesis and Pharmacogenetics Side of Renin-Angiotensin-System in COVID-19. Genes (Basel). 2020 Sep;11(9):1–17.
79. Sriram K, Insel PA. Risks of ACE Inhibitor and ARB Usage in COVID-19: Evaluating the Evidence. Clin Pharmacol Ther. 2020 Aug;108(2):236–241.
80. Saiz-Rodríguez M, Belmonte C, Román M, Ochoa D, Jiang-Zheng C, Koller D, et al. Effect of ABCB1 C3435T Polymorphism on Pharmacokinetics of Antipsychotics and Antidepressants. Basic Clin Pharmacol Toxicol. 2018 Oct;123(4):474–485.
81. Mahmoudpour SH, Leusink M, Putten L Van Der, Terreehorst I, Asselbergs FW, De Boer A, et al. Pharmacogenetics of ACE Inhibitor-Induced Angioedema and Cough: A Systematic Review and Meta-Analysis. Pharmacogenomics. 2013 Feb;14(3):249–260.
82. Yang YY, Lin HC, Lin MW, Chu CJ, Lee FY, Hou MC, et al. Identification of Diuretic Non-Responders with Poor Long-Term Clinical Outcomes: A 1-Year Follow-Up of 176 Non-Azotaemic Cirrhotic Patients with Moderate Ascites. Clin Sci [Internet]. 2011 Aug 9;121(11):509–521. https://doi.org/10.1042/CS20110018.
83. Badary OA. Pharmacogenomics and COVID-19: Clinical Implications of Human Genome Interactions with Repurposed Drugs. Pharmacogenomics J. 2021 Jun; 21(3):275–284.
84. Lin CM, Cooles FA, Isaacs JD. Basic Mechanisms of JAK Inhibition. Mediterr J Rheumatol. 2020 Jun;31(Suppl 1):100–1404.
85. Khoury HJ, Langston AA, Kota VK, Wilkinson JA, Pusic I, Jillella A, et al. Ruxolitinib: A Steroid Sparing Agent in Chronic Graft-Versus-Host Disease. Bone Marrow Transplant. 2018 Jul;53(7):826–831.
86. Yee SW, Nguyen AN, Brown C, Savic RM, Zhang Y, Castro RA, et al. Reduced renal Clearance of Cefotaxime in Asians with a Low-Frequency Polymorphism of OAT3 (SLC22A8). J Pharm Sci. 2013 Sep;102(9):3451–3457.

Index

R

S

T

U

V

W

X

Z

For Product Safety Concerns and Information please contact our
EU representative GPSR@taylorandfrancis.com Taylor & Francis
Verlag GmbH, Kaufingerstraße 24, 80331 München, Germany